Gerry –
It's all about the energy
Here to help with [illegible]
Mike

The Green Guide to Power

Thinking Outside the Grid

Ronald H. Bowman, Jr.

For further information, articles, and speaking engagements, contact Ronald H. Bowman, Jr. at: Ronbowman123@aol.com

Edited By: Maureen Gralton Bowman

Bowman, Ronald H Jr. 1960
The Green Guild to Power- Thinking Outside the Grid

Copyright @ 2008 Ronald H. Bowman, Jr.
All rights reserved.

ISBN: 1-4392-0769-0
ISBN: 13: 9781439207697

Visit www.booksurge.com to order additional copies.

Printed in the United States of America

Dedication

This book is dedicated to Connor and Ceara.
What you are I can't give you.
Who you are, no one can take away,
Only you can give it away.
You are the best!

Acknowledgements

The greatest contribution for the following is the unyielding support and work from my editor and wife Maureen. Technically, as a copywriter, she has made this book readable and minimized the "Bowmanese" vernacular I take for granted. She has been supportive through our large challenges and tolerant of the small ones. You are outstanding. I respect your work tremendously. I love you madly. Dick Gralton would be proud of you.

Good friend, John Krush, has provided the intellectual guidelines for my multiple theses. I try them out on John first. John helps me draw inside the lines. Thank you for your time and counsel.

Peter Gross, thank you for your sanity checks, your friendship and your courage.

Thanks to Connor and Ceara, my co-writers, you have kept me on course as to not make the copy too technical. Their curiosity inspired their tireless efforts in research. Thanks to Levin for almost helping. I got your back, man.

Emily van Buitenen, thanks for working tirelessly and continuously to keep this project moving forward. Thank you for all your support.

Vince Rothemich for his wisdom.

Richard Keiler, thank you for the support and encouragement to tell these stories. You're a treasure.

To Bob Bodey for Data support and guidance.

To Steve Lane for his early vision and execution.

Table of Contents

PREFACE

This book is meant to inspire critical thought and discussion about the importance of energy creation and distribution globally. It is intended to inform and guide. The speed of information is equal to the speed and context of disinformation or sound bites.

For users, help is here for decentralized and centralized solutions that can be leveled and weighed. The "how much," "how long," and "how reliable" is now available in deliverable form. A tsunami of energy creation bad news is on the horizon. A perfect storm that invites the "plan now or pay later" challenge like never before.

I do not minimize the importance of conservation efforts, but new alternative energy creation sources are in dire need. Both the United States and the world at large are looking for new energy sources as soon as possible. Currently, demand out paces supply in most parts of the world from 3% to 25% per annum, but transportation CO2 emissions account for only 14% of the global footprint and energy creation is 25% (3% growth for 90% and 25% for 10%). Transportation is more visible and generation is more serious. At the moment, the process of negotiating users off the grid during predictably high energy periods or paying usurious rates to peak power plants make the network dependable but not reliable.

It is a financial control. Eventually to wash your clothes midday may cost you 31 cents per KWH or 12 cents if you do it before you go to bed. Unfortunately, the current system certainly is not cost effective as well. There are economic, resource and ecological time bombs waiting to go off if this generation does not have the ability, willingness and, most of all, courage to make better decisions in the short term for the long term lifestyle and economic sustainability we have grown accustomed to. The US needs to build 60,000 megawatts of renewable power to meet the existing velocity of demand and the inevitable and realistic decommissioning of currently operating central plants. This may equate to 10,000 new facilities or more. Independent power providers and corporate users will need to participate in a short term spend of over $65 Billion dollars to satisfy new and short term needs. This is budgeting what we know and not what we don't, clearly, abnormal changes upwards will create more pressure to create more energy, renewable or otherwise. Energy and potable water resource creation and management will be the short term "Holy Grail" for sustainable societies and growth.

The immediacy of these challenges is finding us. This can not be overstated by any means or fashion. It's as if we have driven into a wall at high speed and come to a dead stop. The inability to access reliable, cost effective and carbon neutral power is affecting individuals, businesses and the planet. Water challenges are target rich. At this time, water purchases are one of the largest annual spends for most municipalities-second only to education. At the same time, creation and distribution issues are not getting better in the near future. This book is dedicated in most part to energy and electrical challenges with solutions. However, as most mission critical users have learned– without water there is no power...

Please read the following with an open mind and realize that a "Man on the Moon" effort or spend, a "D-Day" coordination and "Desert Storm" execution will be necessary to make it successful. A collective effort is required by both public and private sectors to alter our path. This will be to say the least, tricky. The public sector needs to incentivize or inspire investment (as they have in Europe) but the markets need to prevail at the end of the day. If ethanol is the government's way of guiding us through the alternative and additional fuel mine field we currently face, most of America would like to change direction. Although "green" is the new red, white and blue, there is no silver bullet to today's challenges. Multiple technologies and financing tools will be needed to be employed to make the solutions practical and usable within reasonable time and expense horizons. The hyperbole title NOPE – "Not on Planet Earth"- reflects a ton of left interest to not drill for additional oil in the United States. I call it hyperbole because transportation only conducts to 12 percent of our overall CO2 emissions. It is, however, very politically correct to be pro environment with most pundits trying to out shout each other on cable news with the last data point they overheard. This book does not have a position politically. NOPE is the other side of "Drill it all."

The "Power Network" in the United States is the most complex and successful "Just in Time" solution for lifestyle, business, entertainment and transportation needs created by man. This network includes 1,000 power companies with 600 generating plants. These include 103 nuclear facilities with a majority of fossil burning and a minority of hydro or renewable plan to, all working. This entire network is "just enough" and "just in time" to make things work.

Currently, our system is divided into regional consortiums which govern their own generation capabilities, maintenance, and cost. After deregulation in 2002, the system was divided into "generating companies" and "wire or billing" services. With profit margins thin for both services, change and improvements keep reliability and cost containment that much more challenging. The Clean Air Act of 2002, Clear Skies Amendment of 2004 and the Energy Security Act of 2007 were landmark pieces of legislation passed and aimed largely at retail and conservation energy issues. Power generation and creation issues are addressed, but the lasting effects of the legislation are more psychological than practical. Like many pieces of landmark legislation, subsequent amendments and changes will meet the market in terms of needs, but it is and was impactful in that change is coming with memorialized forethought to the years 2010, 2020 and 2030.

This is not a "good guy-bad guy" story of the monopolies taking advantage of the end user. It's not a story of what happened when the US was drilling for water in Saudi Arabia, struck oil and the subsequent impact that has had on the Middle East. Our reliance on oil has influenced policies and politics in both the US and the world at large.

The following is a brief history of power creation in America. This book will address the distribution of power to the urban and suburban footprints, as well as the environmental challenges we share globally. Realize that pre-deregulation, the placement of generating stations, substations, and transmission lines were not always built with the best strategic planning. I will discuss the "centralized" and "decentralized" energy creation choices that we now have and the environmental relevance for each. I will cover:

How we got here– a more specific history of power creation in the US. This will include an objective review of Regulated and Deregulated power, and what both have meant to power generation, transmission and distribution.

What is here? What are the choices and challenges of centralized and decentralized power creation, distribution with the economic and ecologic nuances of the private sector, public sector and the user?

How to move forward-One size does not fit all.

I will articulate today's legislative guidelines, and new goals and laws for such documents as the Kyoto protocol. I will discuss the supply and demand, incentives as well as legislative challenges.

I have a required discussion and analysis of growth and availability of:

Hydro power-History and availability- irrigation and distribution impact of potable water needs for human agriculture. A majority of dam's create energy.

Fossil power-Coal/Clean Coal/Gas/Oil- Fossil power is responsible for approximately 70% of the total creation of energy.

Nuclear power-103 plants and growing in the US. Currently, there are 440 operating worldwide.

Alternative power- Geothermal, solar, wave, tidal, shale oil, sand oil, wind, bio mass, bio fuel etc. At this time, alternative power represents less than 2% of power creation. Data points are fluid and sketchy, mostly self serving to the authors. When did scientists become economists?

I will discuss the environmental impacts of "existing" sources, transmission, and distribution.

I will discuss environmental impacts if "other" sources, transmission and distribution.

Extraordinary efforts are being made for various reasons to conserve different uses of power. The public sector and private sector are engaged in the challenges and creating solutions in a "one size does NOT fit all" matrix. We need to work together to provide cost efficient and reliable power, measuring usage and waste as one of our top priorities. One could take the position that this is too little effort, a bit too late, but better late than never! Our government cannot create incentives, motivate conservation, or create power without credible data points. We are at a "measure twice and cut once" stage of legislation creation, but the data creation is almost entirely voluntary. By the time the collective data has been collected, collated and analyzed, it may well be less relevant. "Energy Star" is the latest and best effort the Department of Energy" (DOE) is using to create relevant data points to address today's concerns and rewards to same. Power creation efforts need to be and are tantamount to conservation efforts. Energy conservation will lead the way and provide the biggest bang for the buck in the short term and has all the feel good components we are looking for. Energy creation is the long term and more impactful solution. New missions and motivations for both will be established from beta testing and data point collecting now underway by the Department of Energy- "Energy Star" buildings also get a very cool "Energy Star" plate to display to fix to the building so tenants will have the "feel good" conservation emblem in their face every day. It will be an attractive marketing tool for landlords, as well. So-called "Green Buildings" are the wave of the future.

1) **Residential** users are converting over to solar and other decentralized energy alternatives. Over 80% of growth has been realized year after year and revenues of over 6,000% over 7 years for many manufacturers. Sustaining these profits is remarkably unlikely. We are now at the "shorting phase" of renewable energy stocks. Unless forward PE's line up with some reality or "mark to reality" check, we will quickly find these stocks in the "Dot -Gone" phenomena of 10 years ago when stocks traded on "eye balls" and "homes passed" or "buildings lit". Renewable energy companies have to make money and turn EBIDA positive to really get market credibility and market sustainability just like deregulated telcom companies did. Other than that, they are a trade. Good companies were destroyed due largely to unrealistic analysts expectations of companies. These analysts were in the business for effectively 20 minutes and were applying the "new math" to story stocks that often built bridges to nowhere and those that did were taken out back and shot because they were conservative in their building and spending. Similarly, today the energy and alternative energy stocks have the same "buzz" and covered by similar "newbies" of energy technology that unfortunately few people understand. There will be winners and losers, but we are clearly at the "excessive exuberance" stage and "you pick em" - dart throwing stage of stock/technology winners and losers.

Commercial office buildings are using the technology aggressively. Today's technology allows variable speed drive motors to monitor and manipulate controls for temperature sensitive parts of the day. There is more thermal storage and peak load shaving, and so on. "Heat Islands," aka cities, are 10 to 15°F higher than urban and up to 20°F higher in Tokyo.

- More emististic perimeter thermal pane glass is being installed in office and residential buildings. This glass is energy efficient in that it absorbs and retards the destructive heating rays of the sun to keep hot air out and cool air in.
- More offices and interior space have parabolic lenses for greater light efficiency. This allows more natural light to penetrate into center core and side core buildings coupled with reballisting.
- Thermal storage and alternative cooling methods are becoming more and more mainstream. Making ice slurry and blowing fans over cool "super slush" reduces peak usage and peak billing of energy.

- Existing lighting is now more efficient due to parabolic lenses, new ballasts and new bulbs.
- Technology or software now manipulates light and power needs in a more cost efficient way designed and driven by function and not form. Technology is actually working for us, not the other way around.
- Lighter- Whiter buildings and parking areas.

2) **Industrial-** natural light penetrates roofs and perimeter more effectively.

- Roofs are white or green to reflect heat or capture ambient cooling or heating.
- More gas and other alternative and decentralized solutions can be used rather than electricity services. Industrial parks are convenient for more efficient heating and cooling.
- Combined heating and power or cogeneration and trigeneration of power and create heat containment for thermal benefits which are widespread.

Our ongoing efforts towards conservation need to be eclipsed by efforts towards generation in the near term. Demand has been outpaced by supply by over 20% annually for certain industries and will continue at an increased velocity going forward. Here, we will identify the macro and micro dynamics of supply and demand of power generation and distribution. I will describe the early benefits and recent shortcomings of existing traditional methods of creating power as well as the economic and short term challenges to improve and maintain these systems.

Most importantly, this will identify the most effective energy creation and conservation methods of procedure, associated day one operating expenses, sustainability and maintenance requirements for all systems. The two fundamental sources of energy creation are "centralized" and "decentralized". The centralized are the incumbent sources we are familiar with and decentralized are often closer to the end user(s).

Centralized Generation-is associated with 400-5,000 megawatts created in our 1,000 energy company infrastructure and plants usually remote to usage. Centralized generation is electronically distributed through a transmission and distribution network and is generally the most dependable, but not the most reliable. The system has incumbent and variable risks, costs and inherent loss of user control.

Decentralized Generation is defined as a system close to the point of creation from 1 KW to 50 megawatts. The benefits are:

1) **Built to suit needs and scaled to unique requirements.**
2) **Reduces or avoids the main transmission grid. It also can be blended often.**
3) **Diverse energy sources reduce risk and catastrophic failure.**
4) **Can provide premium or computer grade power when combined with UPS.**
5) **Appropriate for renewable technologies because they can be located closer to the user and installed for an "as needed" basis.**

The following is a brief history of the energy creation and conservation movements (past and present) that meet the demand for growth in the urban, industrial and rural markets.

What inspires today's unique interest in power generation and power distribution is recent deregulation of the utility, The Clear Air Act of 2002, The Clear Skies Amendment of 2005, The Energy Security Act of 2007, The Kyoto Protocol, CO_2 emission concerns, and the "Big Six" air particulants associated with global warming (CO2, SF6, HCF, NOX, PCF's and CH4).

What is different from the fuel embargo and fuel shortage days and fuel rage during the 1970's (1973–1978) when we bought fuel by the odd or even number in your license plate? What we're experiencing now is a confluence of extraordinary power consumption driven by technology at the "small office" home and home offices (SOHO's) and the recognition of finite fossil fuels. Supplies are short and ecological impacts are multiplied by electrical devices at the office and home with associated cooling that have a higher power load per sq ft. There are 300 million Americans and 138 million of these Americans are in the work place. There are 250 million computers fixed in place, and approximately half those amounts are laptops. Of the 162 million people not in the work place, 80 million are students or kids with multiple PDA's: games, phones, I - pods etc. The point is that most working adults and students have multiple electrical devices working congruently with Moore's Law of bandwidth growth that needs to be supported by today's and tomorrow's power grid. To expand on that thought, even if there was a commercially deployable electronic car today, our power grid could not support wide spread usage and "charging."

Data center or mission critical facilities of the 1990's allowed the telecom deregulation and the 20% per annum increase of power that sector gave us across various enterprise businesses. In Chapter 12, I will memorialize the initiatives of the Federal, State, and Local governments driven by corporate governance as well as a remarkably concerned and motivated general public. I will identify best practices of the inside plant-ISP and I will illustrate best practices of the outside plant- OSP, realizing that the first urban and industrial footprints in the US have changed significantly since the design and installation of the generating, transmission and distribution infrastructure improvements built between forty to fifty years ago.

There are no bad conservation measures. There are some alternatives that have better ecological benefits than others. Some efforts actually create more waste or effort in the interest of reduced emissions and some alternatives are just technologically too commercially early or cost inefficient in their individual "well to wheel" cycle. It will be necessary to track and validate the data points and costs to create, distribute, use and bill for energy. For example, the early hydrogen power was created by the US Government for the Apollo space program at $600,000 per KW. It is now approximately $2,500- $4,500 per KW to create. This is still too high to be commercially viable for some. The challenge to manage waste energy created by separating hydrogen molecules from water molecules in many but not all cases is cost inefficient. In other words, it often takes more energy or waste to make it than to realize the ecological benefits of using it. Hydrogen from natural gas is more commercially viable in urban environments for base loads and load shedding programs. With the help of one time capital dollars and recurring operating dollars, the hydrogen fuel cell is making more "cents" for regions where the cost per kilowatt hour is 14-18 cents and the price of natural gas is under $14. The Carbon Credit conversation and financial benefits from Regional Greenhouse Gas Institute (RGGI), Chicago Carbon Exchange (CCX) or others will only increase a total cost of ownership (TCO) model.

There is a cost or effort to create all forms of energy. Exclusive of reliability, hydro, geo-thermal, wind, solar, bio fuels, fuel cells, cogeneration, etc. all have cost benefits and end products that need to be leveled and weighed. Most data points available are vendor enhanced and most data points by end users are proprietary or non transferable due to field conditions, climate or location. The trick is how to control the variables to ensure to the user's benefit. In saying that, it is pretty obvious that I have a personal bias for bio waste (methane), tidal, algae and jatopha for renewable energy sources.

I will discuss the state of power generation globally. The CO_2 emissions condition we find ourselves in is worldwide. India and China have extraordinary demand needs and challenging environmental issues to satisfy same. The U.K. and Continental Europe are split on the future of power. Germany is decommissioning all of their nuclear facilities by 2021, while France is creating more nuclear energy within the EU than any other country and exporting most of it. The licensing application and new construction of nuclear facilities is up 50% in Europe. Nuclear power was once viewed as an environmental mistake and is now being embraced appropriately as a green solution. Other than the 103 nuclear generators in the US, there are 337 globally. At this time, 28 new facilities are under construction, 62 planned and another 162 proposed. At an estimated $5 billion to build, and $10 billion to decommission, the economic impact of nuclear power is extraordinary. The impact of water drought to manipulate turbines and cooling system has shut down nuclear plants across the world. The importance of water for hydro energy creation and water's value to maintain nuclear power can never be minimized.

Depending on your non partisan brand of global warming, most people agree that CO_2 emissions and other pollutants are warming the planet to unnatural levels. Higher CO_2 levels have created a gaseous blanket around the earth, and warmer water temperatures are acting as a catalyst to expand these water molecules and raise water levels. Cold water molecules are smaller than warm water molecules. I'm sure male readers will identity with this. Those facts, coupled with the melting of the polar ice caps, water levels at our shore lines are becoming a critical issue to cities and their power infrastructure.

The sky is not falling. A common sense view of the scientific data points is required. If not, a dot-com spend will yield marginal benefits. By way of example, some believe that full compliance by the USA to the Kyoto protocol will cost $15 trillion dollars and yield 5 to 7 years net benefits.

Most CO2 emissions do not come from cars or smoking, but rather from power generation (concrete and steel production are close behind). However, spontaneous combustion of coal seems to amount to 2-4% of the annual CO2 emissions worldwide. This means that fires burn 24 × 7 in the coal or gas rich parts of the world (caves, mines, ect.) unabated, creating huge CO2 contributions on a daily basis. Coal fires in China amount to 109 million tons of coal a year emitting 200 million tons of carbon monoxide. This is simply too much for the planet to sustain. Perhaps we should put out some of these fires. Since there really is no profit or business related to putting out fires or spontaneous combustions, these fires will likely continue until the emissions are measured and a benefit or profit is framed or articulated to motivate the extinguishing of the fires.

If we are sincerely interested in global warming or reducing the newly created thermal blanket that surrounds the earth, we need to take care of it now. The warmer ocean is rising .36 inches per year, or 7 inches over 20 years and flooding 45 miles of coastal regions. Before long, we could lose vast areas of our coastal lands. Also, these increased water levels are creating not more, but more powerful hurricanes and tropical storms, power generation will have the biggest impact improving or destroying the atmospheric and water tables. Half the US population lives within 100 miles from the coasts, and over 100 million people live within 50 miles of same in the urban environments and coastal cities and towns. Data points are sketchy about coastal waters. In the 1980's, it was to rise 2.5 feet, in the 90's it was 2 feet and now it is 1 foot over the next 30 years. There is cherry picking of Data by scientist turned economist turned pundits turned alarmists.

How did we get here? Quite simply, the United States population grew by 400% over 100 years. The U.S. population is expected to double by 2050, power generation is not expected to keep pace with the population. In fact, power demand is out-pacing power supply by 21-25% depending on whose data points you buy into. Our power needs will double by 2040-2050, and coal, our large generator source is no longer environmentally viable. The price of short term coal contracts have increased 60% to 100% over the past two years due to increased demand and limited supply. Depending on the quality of the coal, the costs have risen over 200 to 300% over the past 7 years. Over 50 requested permits for new coal facilities have been denied over the past two years due to ecological concerns- 10 in Texas alone operates with 7% of power provided by nuclear sources. Texas, by the way, creates more wind energy than any other state in the US. This is appropriate, if the state of Texas were a country they would be the 6th largest CO_2 contributor in the world due largely to coal burning energy creation. The utility's long term coal contracts have not yet expired for the most part and many of the existing billing rates have not realized the increases in the billable cost per kilowatt hour. When they do, the real inflationary impact on the home, commercial and industrial operating expenses will be immediate and measurable. The electrical or utility bill could easily double. That will go from $200-$300 dollars per month to $500 to $700 per month. This is equivalent to a large car payment, a lot of trips to Home Depot or, funds for retirement or education.

Realize that most of our 103 nuclear facilities were built within 10 years of each other--only 3 have been decommissioned. The facilities were granted 20-30 year licenses. Most were extended to 60 years, often with suggested improvements. We cannot just extend their useful life based on need. The useful life of many of these plants will end as swiftly as they came on the generating market. Recently, the Indian Point nuclear facility in New York's second and third renewals have been declined. Reactor 1 was closed in 1974, and now reactors 2 and 3 are due to close in 2013 and 2015, respectively. If they cannot renew, that will leave a huge hole of approximately 2,000 megawatts in the power creation supply side in the Northeast.

The fossil solutions of gas and coal are approximately 70% of the national grid. Natural coal is abundant and if not treated properly, a dirty and fast source of energy. Utilities use 50% coal and 19% gas, 19% nuclear and 12% other. The cost of natural gas has gone up 30% in the past 3 months and 70% of New York City runs on natural gas energy sources. The US, like TXU, are weaning themselves off coal to gas and the cost per kilowatt-hour reflects that. They are not alone; California and PG&E are eco friendly utilities that are paying the price to get off coal with the nuclear facilities almost maxed out. Renewable sources, like wind, have storage challenges that are unmet and tying into the grid is an unknown cost and challenge.

The hydro facilities are the most eco friendly sources of the traditional and widespread power creation methods. They create 9% to 10% of the grid and the dams are good a 24/7 source, while waves or wind have obvious and natural dips (pumps are used for some reservoirs to mitigate low water flow the same can be done for tides). Hydro as a source of fuel is low cost and reliable. The dams are fairly central to the United States, but the water movement is lower and slower due to the irrigation and growing national population. The need for agriculture and human needs are conflicting at an accelerating rate with energy creation demands put on our natural and national resources. Many of these large rivers look like marshes or creeks by the time they reach the southern states and irrigation and the major or micro dams have done their jobs.

As of now, the national grid is made up of the several generating sources. The largest source will be significantly reduced in the short term. The US is currently exporting coal due to the rejected permits for new coal facilities as well as market demand and high market prices. The nuclear sources will commence de-commissioning faster than the commissioning at twice the rate, and we cannot just make more water in the mid states for new hydro power. The "other" category is the one to keep your eye on for central and more importantly decentralized solutions to new power generation and distribution. "Other," needs to grow at a disproportionate velocity of speed for generation and distribution. Remember the distribution!

Nuclear	19%
Hydro	8%–10%
Fossil-Coal	50%
Gas	19%
Other	2%

When I last did this analysis, perhaps eight to ten years ago, generation was split almost equally between fossil fuel, nuclear and hydro powers. Now, where do we go from here? Natural gas is 1/3 of fossil and a likely growing source of "other" sources of meaning and reliable power to be expanded but also up 35% in costs over the past 3 months. The collective 2% of "other" sources of power need to be grown in the market place with state of federal subsidies or incentives, loan guarantees, bonds or grants if need be to satisfy the growing demand for the following distribution. "Other" costs inclusive of one time and opportunity cost more per kilowatt hour than hydro, nuclear or fossil fuels.

40%   **Residential**
30%   **Industrial**
30%   **Commercial**

We should embrace the idea that burning coal, shale or sand are viable and abundant, but have an unfavorable carbon footprint. Capturing and reusing carbon is and needs to be a high priority in energy creation. Canada has the second largest sand footprint and the US has 30% of the shale oil. There is a sand oil "footprint" the size of Florida within Alberta, Canada. The US is rich in natural gas by global standards, but still has finite resources. Fossil fuels are 75% more energy carbohydrate rich than biomass or man made energy. That means "this is the good stuff." For athletes, think of fossil fuel as "Red Bull" or dense energy supply and biomass is like "Gatorade", a less dense energy supply, but does the job. For those readers in a twelve step program who enjoy living on the edge, nuclear power is the cocaine, but sustainable and fossil fuels and renewables are like light beer. On the other hand, what a waste of such a precious and carbohydrate rich substance! We are literally burning what nature took hundreds of millions of years to make in over 1,000 times it took to make it. In the timeline of life, if a 400 sheet roll of toilet paper represents the life span of the earth, man has been burning fuel past the industrial revolution for about one finger length of one sheet and we have enough fossil storage left for about two fingers.

I will discuss in detail the state of technology for the following decentralized energy alternatives, their maturity to the market place, reliability, financial feasibility, interoperability with incumbent utility and ecological impact.

- Wave- The vertical manipulation, capturing, storing and distributing of energy created by waves.
- Solar- The capturing of the suns rays by heat and storing and distributing of same.
- Bio-mass- The capturing of energy by the decomposition of burning of organic materials. (Storing and distributing of same. Waste, trees, grass, corn, algae, jatropha etc.)
- Bio-waste- The capturing of methane gas from decomposition and fermentation of human and animal waste.
- Geo-thermal –The capturing of the earth's core energy, heat via gases (nitrogen), steam and hot water through the earth's surface and vents in the ocean floor. The storage and distribution of same.
- Tidal- The hydro energy creation of power by water capture and manipulation via the lunar tidal activity

near shore lines. The storage and distribution of same.

- Wind- The creation of energy by manipulating the wind via wind mills or blades, and storage and distribution of same.
- Fuel Cell (Hydrogen)- The capturing of energy by the chemical and molecular transformation of hydrogen from gas or water and the release of power when exposed to the chemical flow in a controlled vacuum.

My concept is for business to no longer depend on the grid completely, or the traditional methods of energy creation or distribution. Alternative or hybrid solutions are now commercially viable and ecologically required. The economics, "Acts of God" and legislative challenges of growing the existing grid are daunting. Public sector efforts (Department of Energy) are concentrated on the measuring of the usage and conservation efforts of users and at utilities. This means we are not even on the solution side of the power paradigm we are faced with currently. (Government and private sector beta programs like "Future Gen, Clean Coal" bring the biomass carbon retention prerequisite of 2007 and now put it off till 2010 or 2012 due to changing technology, RFI's and doubling of proposed cap ex from 950 M to 1.8B....with a "B." Users need to take the future of power creation and energy reduction into their own hands, not only to be good corporate citizens, but to ensure the long term viability of its operations and protect their brand and as well as their market share.)

Even if we replace coal plants in gas fired facilities, the substations that need to grow to accommodate the former rural acres are now in thriving substation environments.

Schools, homes, shopping centers, and the like have the high voltage transmission lines and generating substations that are land locked and will likely get pushed back from the "not in my backyard" (NIMBY) crowd to grow substations or improve or add high voltage power lines. The NIMBY crowds efforts are often eclipsed by "Build absolutely nothing anywhere near anything" (BANANA crowd). That means the critical services will take the longest distance between two points (for improvements or new infrastructure) to satisfy the deregulated utility needs and given to a low bidder to implement!

My guess is that these programs will generally become too costly to build and maintain. Options?

Energy creation methods that are decentralized or created closer to the "need" with less equipment and shorter distances to maintain, will have greater reliability, dependability, and maintainability. The corporate user as well as each individual person, will all benefit from cost effective and carbon friendly energy. These facilities could be build out and way from 100 year flood planes they are currently in. That means 10 years of data memorialized what we call "100 year events." Cases of flood and tidal surge are up. Approximately 30% of all losses from Acts of God come from forms of flooding. The more people, and the more businesses in flood zones, equals more damage.

I have written this book with the energy solutions and resources management wisdom accumulated over 25 years. Data centers and large corporate users have tantamount needs of small cities.

This book is meant to teach and guide without the media drama.

Chapter 1:

History of Power in the United States-Green is the new Red, White, and Blue

The power or energy creation system in the United States can be considered one of the most sophisticated contributions by mankind. Early power users in the US were almost exclusively built and distributed to satisfy the urban or densely populated areas. This power was uniquely located near port cites or close to railroads and thoroughfares to satisfy the concentric circles of need associated with commerce.

The following is an abbreviated timeline of energy creation and points of energy interest highlighted by "Mother Jones":

- As early as 1748, a commercial coal mine was opened in the state of Virginia to satisfy the power needs of colonial America.
- Another extraordinary step forward came with the invention of the battery by Alessandro Volta in the year 1800.
- In 1815, natural gas was discovered near the Camorra River in Pennsylvania.

- The First electronic locomotive was created by Robert Davidson in 1842.
- The year 1859 was marked by the foundation of Standard Oil by John D. Rockefeller.
- Thomas Edison created the light bulb in the year 1870, bringing electric light to homes and businesses alike.
- New York City founded its first coal mined power plant in the year 1882. The city was illuminated by coal. (Now 70% gas.)
- In 1908, Henry Ford's horseless carriage was powered by both gas and ethanol.
- Thomas Edison created the alkaline battery in the year 1910.
- In 1911, the Supreme Court broke up Rockefellers Standard oil.
- 1930 – GM, Firestone, and Standard Oil buy up electric trams or trains to replace them with buses. Good for the Auto Production, tires, and fuel.
- 1954 – Bell Labs creates solar "PV" cell.
- 1956 – Federal construction of Interstate highways connects the county...cost-$129 Billion. Here come the trucks and cars! (Abe Lincoln funded the first interstate roads to encourage interstate trade.)
- 1957 – First full scale nuclear power plant in use at Shippingport, PA. (A super sized science project.)
- 1964 – Glen Canyon Dam built to supply energy to southwest.
- 1973 – Oil Embargo.
- 1978 - 79 – Iranian Revolution sparks oil conservation and increase of prices. Presidents Nixon, Carter, Ford, Reagan, Bush I, Clinton, Bush II vow to be energy independent.

- 1979 – Three mile Island Nuclear Power Incident
- 1987 – Yucca Mountain, Nevada designated as nuclear storage facility. Over budget by over 100% at $19 Billion. As of today, it is still not open. The 17 volume 8,600 pg application was received by truck on June 3, 2008. Approvals should be in place by March 2013. Opens in March 2017.
- 1997 – GM rolls out EV1-All electric car, very successful.
- 2003 – GM destroys all but a few EV1 cars, also very successful.
- 2005 – Congress triples ethanol production goals (this is a very large bet for energy crops).
- 2007 – Americas Climate Security Act (S2191) establishes greenhouse gas caps to trade. Empowers EPA to regulate.
- 2008 – Congress and lobbyists are exposed as sources of misguided interests in corn legislation. Cost of meats, poultry and dry goods rise with corn shortages and corn syrup products that are part of the food chain.
- 2008 – Carbon Sequestration legislation induced for Fossil fuel generating plants (retrofits and new pre and post combustion).

The deregulation of the utilities created more regional dependencies which brought generation and transmission changes, price impacts, reserve shortages and less reliability of the networks.

Today, main NOCC's (network operating control centers) prepare loads forecast two hours ahead. Energy is made the instant it is needed, and is used the instant it is made.

It cannot be stored. Data is stored to anticipate weather changes and holidays. (When people get up and go to bed.) When there is a ball game, election and other living habits, however human and weather changes are always a fixed variable.

Electricity is a $24 \times 7 \times 365$ need. In coal fuel power plants, approximately 3 days of need can be stored immediately adjacent to the generation facility. Briefly, coal is the end product of our ecosystem's waste products. It is the result of years of development and significant pressure on the plant remains or vegetation saved by water and soil from oxidation. Coal is combustible and is made up of primarily carbon (along with sulfur). Not only is coal the largest source of power creation (energy) worldwide, but also the largest contributor of carbon dioxide emissions – just ahead of petroleum products. This is not necessarily a well known fact. Coal contributes about twice as much carbon emissions as natural gas. It is a "dirty fuel." There are six commercially recognizable types of coal. Coal also adds to approximately $60 Billion dollars a year to the US economy.

They are:

- Peat – used in Ireland and Finland
- Lignite – Brown coal, low grade, stones used as art. (Common source of "Dirty Thirty" coal fueled generating plants.)
- Sub-bituminous – Fuel coal to make steam
- Bituminous – Black coal – dense – makes steam or coke
- Anthracite – Hard, glossy, black used to heat commercial buildings
- Graphite – Hard to ignite, used in pencils

The world consumes about 6.5 billion tons of coal annually. Most of that consumption, or 75-80% goes towards making power. China produces 2.5 billion tons. 83% of China's power or electricity comes from coal. The U.S. uses 1.2 billion tons of coal, and 93% of that goes towards making energy. The world production of coal is approximately 6.5 billion tons.

Collectively, fossil fuel sources of energy like coal are remarkably inefficient, and the US has approximately 600 power plants supplying 50% of demand. Approximately 70% of coal energy is lost in the by product or heat. "Grandfathered" or old coal generating plants are often 80% inefficient. Then appreciate that from a transmission line to a server in a data center; that is another 65–75% loss through stepping up or down of voltages, the rectification from AC to DC to AC…this end product of energy is very, very valuable to run one 100 watt computer 24 hours a day, 7 days a week, 365 days a year. It takes 966 pounds of coal! That means 30% of coal makes it to energy and 30% of energy makes it to the demand source or about a 90% inefficiency model. This is worse in China where they are three times more inefficient in their energy creation of coal. While China was once able to produce all the petroleum they needed, they will need to import 75% of their needs by the year 2025.

At current rates of consumption, we have approximately 300 years of shared use of coal or 120–175 years of exclusive use of coal are stored on site. It is crushed and elevated seven stories at 100 tons per minute on the conveyor belt to be crushed again into a fine powder-like substance where it is mixed with air and blown into a furnace of multiple burners. Over 10,000 tons a day can be burned at a single furnace. This is equivalent to train 130 cars in one day

Almost instantly, the purified water in the tubes around the boiler is turned to steam, which is directed towards the blades of a turbine of tremendous pressure. This is where steam becomes electrical energy by turning the shaft of the turbine at 60 revolutions per second. Turbine and generators turn at the same rotation making mechanical energy electrical energy.

The power goes out to the transformers where the voltages are stepped up for more economical transmission of power. It is then converted to a stepped down transmitted substation and the voltage is reduced for carrier manipulation and usage.

Similarly, natural gas is a "just in time" solution for creating energy. Natural gas was discovered in 1815 on the Camorra River, near the Appalachian Mountains while drilling for salt in "brine" wells. Early settlers called the gas the "wild sprint" in Ohio, West Virginia and Pennsylvania. That gas was drilled thinned and manipulated. In 1840, in Centerville Pennsylvania, the gas was used to boil the brine and became the first commercial application of gas. By 1900, natural gas was commercially deployed from thousands of feet below the earth.

"Sub-surface geology" was the first study of the earth which was born out of searching for gas. Following the better intelligence gathering for drilling and controlled usage, distribution of gas through a vast network of pipelines would provide gas to parts of the U.S. not rich in gas resources. The "war lines" were built in a North/South network to provide energy during World War II to protect our energy resources. They were installed before building codes were enforced or sensitivities towards earth's movement were discovered. Many of the "war lines," earth and pipes have shifted over time. Currently, the US has over 75,000 well spot locations distributing over 350,000 miles of transmission lines and three times that distance in distribution lines.

As well, these wells leave the fields to central pumping stations where a constant pressure can be established and the distribution or transmission routing controlled by compressor stations. Pressure moves the gas to other regulatory stations for downstream distribution. Every last cubic foot is measured for distribution and sale. Line loss is considered in transmission due to theft, leaks and pressure drops. Gas from Louisiana is billed at 104-105% of demand to make up for the unfortunate line loss. Transmission lines are reduced to transmission lines at the "city gates." The role of flow is based on demand and controlled through the NOCC "just in time" considering time of day, weather changes, and other variables.

"Hydro" energy is the manipulation of flowing water to provide 24 × 7 × 365 low cost energy in our 700 existing dams. Our rivers in the Midwest provide a good deal of the nation's "hydro" power to the central states. Micro dams are located throughout the country. The turbines are turned by the waters pressure mechanically and with the generator, the energy in transferred to electricity. The energy is transmitted to transmission lines and stepped up to lower cost high voltage transmission lines and stepped down for local manipulation and usage.

Nuclear energy is created by uranium neutrons splitting other atoms, bouncing off each other and disintegrating. This disintegration reduces a liquid and creates steam. The steam will drive turbines and mechanically create power via a generator as with hydro and fossil fuels. Early nuclear energy in the U.S. was akin to a super sized science project. Due to the remarkably low cost of fossil fuel creation and usage as well as the long time and high expense of the nuclear power, the first nuclear plant was a "gift" to the world as an example of high cost power plants. Early on, we used the radioactive crops we fertilized for livestock, human consumption and agricultural development. Old documentary films (black and white) show engineers handling radioactive materials freely! Contrary to conventional wisdom, the actual reactor is relatively small and deep inside the cavernous cones or domes we see at a distance. The cylindrical cone channels the steam and air away from the asset. The steel and concrete structures were designed and built to withstand the blunt and burning force of an assault of a 757, the largest aircraft produced or conceived at the time. New reinforcements are designed to consider a 777 aircraft with 1.2 meter or 4 feet thick walls with cylinders as high as 46.7 meters and 45.0 meters wide. By the way, driving an aircraft that size into the openings would be like threading a needle at 200 mph and if the aircraft hit the exterior it would disintegrate or evaporate instantly. Catastrophic damage from the outside is very unlikely, contrary to popular thought.

The operating costs of generating electricity from nuclear facilities are not much different from the costs to operate coal or gas fired facilities, approximately 3 cents pKWH, but the capital costs are significant at $5 billion to commission and $10 billion to decommission a facility. It is always important to consider the lengths of time to build a facility- 13-25 years. The monopolistic regulated world of power generation made it easier to future proof supply overbuilding demand with no fear of competition. Now, with competition comes the relevance of spend of $5 billion and waiting 15 years for power while fossil fuels are faster and less expensive to build, one can see the historic path of least resistance. This is largely why we have not employed more nuclear energy in the United States.

With today's collective energy uncertainties and the traditional sources of energy being depleted, we must focus on the forces of alternative and decentralized power. The world of business, lifestyle and transportation has charged from a fringe or "cocktail party" concern to a silo or focus of corporations and overwhelming social responsibility. Energy conservation and energy creation are now swiftly becoming part of good corporate governance with dedicated and well compensated personnel.

We are now at a point where projects that may not make sense at conception are being subsidized by energy cooperatives to effectively shed load off the grid and utilities. This is done by paying users to "self help" themselves with decentralized energy creation solutions so the generating companies and transmission companies can differ capital projects to satisfy new needs. In other words, it pays to keep legacy or new users off the grid rather than build new infrastructure to satisfy growing organic or anomolic needs. With energy reservation or "just in case" levels at an all time low following deregulation, bad weather or regional epic events leave our "just in time" grid exposed to a very dangerous level. This is not where we should be at this time.

Natural gas actually has no or little odor so we add ammonia or other noticeable additives to smell the gas since the leakage can cause catastrophic damage. Natural gas is described as a BLEVE (boiling, liquid, expanding, vapor, and explosion) by the NFPA. The gas is not only an accelerant, but also an explosive.

As well, these wells leave the fields to central pumping stations where a constant pressure can be established and distribution or transmission routing controlled by compression stations. Pressure moves the gas to other regulating stations for down stream distribution. Every last cubic foot is measured for distribution and sale. Transmission lines are reduced to distribution lines at the "City Gates." The role of flow is based on demand and controlled through the NOCC "Just in Time" considering time of day, weather changes and other variables.

Chapter 2:

Generating Sources-Thinking Outside the Grid

Our national network of deregulated generation and transmission is made up of a series of traditional and alternative power sources to feed the "Grid." These sources satisfy our insatiable appetite for electrical power.

The sources have various "supply and demand" criteria for the energy produced. Hydro energy requires water movement to move the multiple geared paddles or blades of the turbine in synch with the generators.

Fossil fuel requires coal or petroleum products to burn. The natural by-product of energy is heat, which creates the steam or hot water to move the blades of the turbine and in synch the generators.

Nuclear energy employs uranium to act as a catalyst for the electrons and neutrons split and create energy. The heat waste product creates steam to move the blades of a turbine and generator.

I mentioned earlier on in the book that the creation, growth, expandability and maintenance of power in the

United States needs to be as meaningful a contribution of society, as landing a space craft on the moon, the coordinated efforts of "D-Day" and the execution of Desert Storm. The man on the moon concept implies a super sized coordinated effort. The Desert Storm coordination indicates a landmark spend for a single purpose. The D-Day landing and Desert Storm invasion require a plethora of smaller and critical efforts or task to be executed perfectly and in coordination for a common good (also requiring a super sized spend). The Desert Storm execution of a 100 hour war with years of strategy and months of tactical planning make it uniquely successful.

Some things just plain fascinate me:

- How aircraft actually take off and land without crashing.
- How bridges are constructed.
- How large ships are made, float and are maintained.
- How toilets flush in NYC or other large urban cities at the same time without negative consequences.
- How just enough power is created and distributed to urban and suburban or rural environments with the considerations of time of day, heat, cooling, weather, season and catastrophic or unplanned breaks in the generating or transmission network.

Imagine some poor soul shoveling coal in a basement to heat a building. From time to time he hears a yell "more coal" or "that's enough." Our national grid is not too dissimilar to this illustration. We need to create just enough power to satisfy the current need, as power cannot be stored in large quantities for long durations.

We incorporate static or constant loads of minimum life safety usage of common area lighting, minimum elevator usage and overall monitoring. We then add loads created "just in time" to meet the needs of a heat wave, causing more or earlier time of day power usage for air conditioning. This illustrates the need for changing power loads in addition to the mean power consumption for an ordinary work day given the time of day, season and time of year.

In the regulated days of power generation we were regulated by the Power Utility Commission (PUC) to keep a 20%-25% reserved consumption capacity. This means the source was bought and stored on site. This was not subject to media hype of shortages or difficulties of getting coal or gas to a generating facility.

The need is the same in the deregulated markets, but only the market conditions drive the supply side for the "just in time" solutions of morning, heat , cooling, holidays, weekends, etc. The coordination and effort to supply just enough power and not too much power (explosive potential) to the correct substations via the right transmission lines to appropriate distribution conductors is a bit overwhelming. If one does not think this daily effort that we take for granted is not one of man's greatest contributions, then what is? Now, if you really want to break a few brain cells or lose some grey matter, incorporate the benefits, shortcomings and challenges of alternative or renewable energy to the grid like wind, wave, solar, bio fuels, bio mass and others. Then place their respective time of day considerations with the existing infrastructure, incorporate the omnipresent pressures and challenges of sags, interruptions or brown outs. For those of us who got 800's on their SAT's, try to sort out the financial impact of selling Independent Power Providers (IPP's) power back to the grid (generally at one third the market rate or a financial loser to sell), charge some users retail for back-up services and wholesale for re-sale of power and 4 layers of prices for peak usage, time of day and size or quality of power. One last thing, remember that when you turn the light on in your bedroom, you pay for one full hour even though you were in there for a moment looking for your sneakers! Yes, the power network is one of the most extraordinary contributions made by man.

Having stated this, it is also the most target rich or flawed network of infrastructure in the world. Waste and deception, (intentional and unintentional) is at epidemic levels. In a paradigm of limited resources, reserves and technology– the paradigm needs to change. Just follow the folly of the regulated power and deregulated power creation, reservation, transmission, distribution operations, reliability and pricing over the past ten years. There were shortcomings in regulated power and shortcomings in deregulated power, but best practices from both incorporated with alternative solutions are timely.

The most significant change from a regulated to deregulated power or energy industry is the fundamental change in philosophy of power supply or generation. In the regulated world, transmission and generation were created and built "just in time" based on earlier discussed needs and protracted studies of consumers habits, time of day, season for example. In the post regulated or deregulated world, transmission and generation are built in a reactive fashion under the guise of supply and demand.

The name of the game in this "just in time" challenge of deregulated power from day one for shareholders is to cut operating expenses in both power generation and transmission. I will concede that in employment for life models like government, education or the former telecommunications, some people just stay too long and no longer add value commensurate with their income. However, in power deregulation, improvements seem to be based on shortages. Improvements are reactive rather than proactive. This philosophy in the "just in time" network creates shortages without boundaries because we create and transmit power in an almost virtual, but regional basis.

The regulated power system had an extraordinary depth of data to draw down from to project loads, plan generating capacity, and allocate the appropriate time to satisfy generation both long term and short term. It takes 5 to 6 years to build most generating facilities and 13 to 15 years to site, permit, build and commission a large dam or nuclear facility. When you add to that another 4 to 6 years to plan, secure rights of way (ROW's), permit and commission transmission infrastructure for intra and inter state improvements then, real and verifiable data points are critical.

In the deregulated world, the laws or requirements for generating and transmission improvements for the "just in time" network are "supply and demand." The most obvious and transparent indicators of not planning or improving effectively are the increases or spikes in prices to the consumer due to shortages of supply or buying power out of region at a premium. The user's recourse to the price increases is at best foul language. This is simply unacceptable. Deregulated power transmission is really not deregulated with "choice." The user is still in a fixed tariff footprint, even if they choose a gas or cogen alternative. The last mile charges still apply. The alternative to "self generate" has legislative and utility penalties or disincentives (unless you use ethanol) in more regions than not, making self help a real challenge of time and money in many instances. The ethanol movement in the US has done great things for corn farmers, corn lobbyists, large agricultural companies and land owners. The cost of land in Iowa has increased 25% over the past 5 months but, currently the land is under water (literally). The concentric circles of under-supply and over-demand of corn and corn products are now catastrophic. The fear of food shortages and the cost increases of corn and corn syrup products are escalating swiftly. A little bit of evidence coupled with media hyped awareness or economic expectations have created extraordinary consequences. Just remember the alleged gas shortages of the 1970's, the sugar shortages of the 1980's, the rubber shortages of the 1990's and the rice shortages of this year!

Many customers in the northeast, southeast and southwest have had their cost per kilo-watt hour increase by over 100% in the past 5 years. These increases are not fully loaded. By that, I mean that the coal and gas generating facilities made approximately 50% of the power in America. Generating companies have LONG term contracts with mining companies to protect their pricing matrix. The cost of coal has increased by over 100% in the past 3 years, and natural gas is up by almost 100% in the past 2 years. This pricing will show up in the generating companies who really count coal as a source of energy.

In the free market of supply and demand, there is either just enough or not enough of anything. Nobody ever got rich with high inventories, or became wealthy in the short term by over- designing or future proofing a facility for 20–30 years with shareholders trying to make the most of their capital in the short term. In the 24 hour news and business media world, we are a quarter driven economy with an eye on guidance from professional analysis. Two to five year forecasts are lost in the fine print, let alone 10 to 15 years strategies. Utilities are inherently strategic and the markets view themselves as tactical entities. They are tactical when they can sort out a new billing system that residents and businesses cannot figure out but must pay to improve their bottom line. Also, they are tactical when a three man job becomes a one man job in the field. (The guy that drives the truck, directs traffic, and lifts him/her self up in a "cherry picker" for repairs.) Fewer people are being trained to multi-task in order to increase efficiencies and the "bottom line."

The California energy crisis is a good example of this kind of supply and demand conundrum. If the regional companies in California had had 20% or greater reserves as they did in the regulated world, capacity could have flowed to save them. If the gas network was sized at the transmission and distribution level by El Paso from the state line into California, the gas could have made it to the generating sources to create energy at even a higher cost. This was another example of not enough power generation reserves, not enough transmission infrastructure to take over uncommon capacity at times of distress in our "just in time" infrastructure. Nothing is for free here. One needs to understand that there is a premium for reliability and quality of power. As mentioned earlier, this cost the user approximately .5 cents in the old model of regulation for "reserves" Given shortages of natural resources and the new and unique concerns with CO_2 emission control, the costs of short term alternative power and the badly needed capital and operating expenses required by our national grid....we need to embrace a short term cost increase of energy and the overall negative GDP drag it will have on our economy. It is the ugly truth. Large users need to self perform or partially perform to relieve some or all of their financial discomfort. Control points for products and services make a positive ecological impact on the economy. If a data center service company's power bill goes up by over 100% over 3 years and energy is 30% of the real expenses, don't you think that cost will be passed along to the user? This is a snapshot of the projected increases and GDP drag on the economy as energy costs increase. Alternatives to mainstream power are required. Everything is on the table. We are turning over every stone to find or tune energy creation and distribution methods that make sense.

The "also ran" or "road kill' of some of the energy creation or generating alternatives being explored or vetted are:

- Magnetic motors- Inventor Dennis Lee claims that motors create 500% efficiency and are effectively "free energy." Dealerships have been sold, but there is no product commercially deployed as of yet.

- Rain Drops- The Atomic Energy Commission uses special plastic to convert falling rain drops into energy. I'm betting on existing dam technology, coupled with pumping reservoirs at night to get peak energy offsets during peak demands.
- Green Algae- We can grow and manipulate algae. Deprived of sulfur and oxygen, algae produces high qualities of hydrogen, and therefore energy.
- Cow Manure- Contained or controlled in an anaerobic digester, bacteria break down (like human waste) to produce methane, which is then trapped and employed to generate electricity. This technology is deployed in California. Chicken manure is used to power prisons currently. Data centers are considering cow manure to power their facilities in part.
- Old Tires- Burning tires in a controlled vacuum creates diesel fuel, combustible gas and steel. The challenge is that it takes an extraordinary amount of tires to create energy and it is not eco friendly in the least.
- Dirty Diapers- Similar to the human waste concept, a British company takes diapers and plastic to create energy. Too much trouble for too little power. Sorting diapers for energy stinks. Enough said.
- Empty Space- Use a vacuum to create energy... NOT!

Algae and manure are economically viable and being deployed for decentralized energy creation currently.

This is not all gloom and doom. With trading credits properly used and conservation and energy creation best practices being employed, we will stay competitive as a country. The government needs to intervene in the short term with incentives and tax credits to inspire change and inspire alternative sources of power. Right now, energy measurement (although required) may be too little too late. We will survive supply delivery shortcomings, but the economic impact is meaningful. Germany gave 100% tax credits to bio fuel creation and within 2 years 5% of the county was on bio fuels!!! Then, the tax credits were taken away, suddenly bio fuel expansion velocity of growth diminished. Like many development quagmires, the greenhouse gas burdens need to be spent responsibly in research, development, and transmission.

More time and resources need to be allocated to decentralized thinking "Outside the Grid."

Like many fossil fuel sources of energy, coal is remarkably inefficient. Approximately 70% of coal energy is lost in the by product of energy...heat! Grandfathered or old coal generating plants are 80% inefficient. The following is an informational list of the "Dirty Thirty" power generating stations in the world. They are all coal. Then, appreciate that from transmission lines to servers in a data center, there is another 65-75% loss through stepping up or stepping down of voltages, the rectification from AC to DC to AC. This end product of energy is very, very valuable given such inefficiencies. To run one 100 watt computer 24 hours a day, 7 days a week and 365 days a year will take 966 pounds of coal!

1. Dirty Thirty – Europe's worst climate polluting power stations

Rank	Power Plant	Country	Fuel	Start of operation	Operator	Relative Emissions[1]	Absolute Emissions[2]
1	Agios Dimitrios	Greece	Lignite	1984-1986, 1997	DEH	1.350	12.4
2	Kardia	Greece	Lignite	1975, 1980-1981	DEH	1.250	8.8
3	Niederaußem	Germany	Lignite	1963-1974, 2002	RWE	1.200	27.4
4	Jänschwalde	Germany	Lignite	1976-1989	Vattenfall	1.200	23.7
5	Frimmersdorf	Germany	Lignite	1957-1970	RWE	1.187	19.3
6	Weisweiler	Germany	Lignite	1955-1975	RWE	1.180	18.8
7	Neurath	Germany	Lignite	1972-1976	RWE	1.150	17.9
8	Turow	Poland	Lignite	1965-1971, 1998-2004	BOT GiE S.A.	1.150	13.0
9	As Pontes	Spain	Lignite	1976-1979	ENDESA	1.150	9.1
10	Boxberg	Germany	Lignite	1979-1980, 2000	Vattenfall	1.100	15.5
11	Belchatow	Poland	Lignite	1982-1988	BOT GiE S.A.	1.090	30.1
12	Prunerov	Czech Republik	Lignite	1967 & 1968	CEZ	1.070	8.9
13	Sines	Portugal	Hard coal	1985-1989	EDP	1.050	8.7
14	Schwarze Pumpe	Germany	Lignite	1997 & 1998	Vattenfall	1.000	12.2
15	Longannet	UK	Hard coal	1972-1973	Scottish Power	970	10.1
16	Lippendorf	Germany	Lignite	1999	Vattenfall	950	12.4
17	Cottam	UK	Hard coal	1969-1970	EDF	940	10.0
18	Rybnik	Poland	Hard coal	1972-1978	EDF	930	8.6
19	Kozienice	Poland	Hard coal	1972-1975, 1978-1979	state owned	915	18.8
20	Scholven	Germany	Hard coal	1968-1979	E.ON	900	10.7
21	West Burton	UK	Hard coal	1967-1968	EDF	900	8.9
22	Fiddlers Ferry	UK	Hard coal & oil	1969-1973	Scottish & Southern	900	8.4
23	Ratcliffe	UK	Hard coal	1968-1970	E.ON	895	7.8
24	Kingsnorth	UK	Hard coal & heavy fuel oil	1970-1973	E.ON	892	8.9
25	Brindisi Sud	Italy	Coal	1991-1993	ENEL	890	14.4
26	Drax	UK	Hard coal	1974-1976, 1984-1986	AES	850	22.8
27	Ferrybridge	UK	Hard coal	1966-1968	Scottish & Southern	840	8.9
28	Großkraftwerk Mannheim	Germany	Hard Coal	1966-1975, 1982 & 1993	RWE, EnBW, MVV	840	7.7
29	Eggborough	UK	Hard coal	1968-1969	British Energy	840	7.6
30	Didcot A & B	UK	Hard coal & gas	1968-1975, 1996-1997	RWE	624	9.5

Table 1.1.: These 30 power plants are the biggest CO_2 emitting power plants in EU25 countries in absolute terms (million tonnes of CO_2 per year). WWF has ranked the 30 biggest emitters according to their relative emissions.

[1] Grams of CO_2 per Kilowatt hour (g CO_2/kWh). Where two plants have the same relative emissions, the plant with the higher absolute emissions (million tonnes CO_2 per year) ranks dirtier.
[2] Annual emissions for the year 2006 in million tonnes of CO_2 (mtCO_2)

Focus 1) Germany's worst climate polluting power stations

Rank	Power Plant	Country	Fuel	Start of operation	Operator	Relative Emissions[1]	Absolute Emissions[2]
3	Niederaußem	Germany	Lignite	1963-1974, 2002	RWE	1.200	27.4
4	Jänschwalde	Germany	Lignite	1976-1989	Vattenfall	1.200	23.7
5	Frimmersdorf	Germany	Lignite	1957-1970	RWE	1.187	19.3
6	Weisweiler	Germany	Lignite	1955-1975	RWE	1.160	18.8
7	Neurath	Germany	Lignite	1972-1976	RWE	1.150	17.9
10	Boxberg	Germany	Lignite	1979-1980, 2000	Vattenfall	1.100	15.5
14	Schwarze Pumpe	Germany	Lignite	1997 & 1998	Vattenfall	1.000	12.2
16	Lippendorf	Germany	Lignite	1999	Vattenfall	950	12.4
20	Scholven	Germany	Hard coal	1968-1979	E.ON	900	10.7
28	Großkraftwerk Mannheim	Germany	Hard coal	1966-1975, 1982 & 1993	RWE, EnBW, MVV	840	7.7

Table 1.2.: Ranking of Germany's biggest emitting power plants according to their level of efficiency

[1] Grams of CO_2 per Kilowatt hour (g CO_2/kWh). Where two plants have the same relative emissions, the plant with the higher absolute emissions (million tonnes CO_2 per year) ranks dirtier.
[2] Annual emissions for the year 2006 in million tonnes of CO_2 (mtCO_2)

Focus 2) The UK's worst climate polluting power stations

Rank	Power Plant	Country	Fuel	Start of operation	Parent Company	Relative Emissions[1]	Absolute Emissions[2]
15	Longannet	UK	Hard coal	1972-1973	Scottish Power	970	10.1
17	Cottam	UK	Hard coal	1969-1970	EDF	940	10.0
21	West Burton	UK	Hard coal	1967-1968	EDF	900	8.9
22	Fiddlers Ferry	UK	Hard coal & oil	1969-1973	Scottish & Southern	900	8.4
23	Ratcliffe	UK	Hard coal	1968-1970	E.ON	895	7.8
24	Kingsnorth	UK	Hard coal & heavy fuel oil	1970-1973	E.ON	892	8.9
26	Drax	UK	Hard coal	1974-1976, 1984-1986	AES	850	22.8
27	Ferrybridge	UK	Hard coal	1966-1968	Scottish & Southern	840	8.9
29	Eggborough	UK	Hard coal	1968-1969	British Energy	840	7.6
30	Didcot A & B	UK	Hard coal & gas	1968-1975, 1996-1997	RWE	624	9.5

Table 1.3.: Ranking of the UK's biggest emitting power plants according to their level of efficiency

[1] Grams of CO_2 per Kilowatt hour (g CO_2/kWh). Where two plants have the same relative emissions, the plant with the higher absolute emissions (million tonnes CO_2 per year) ranks dirtier.
[2] Annual emissions for the year 2006 in million tonnes of CO_2 (mtCO_2)

Focus 3) Poland's worst climate polluting power stations

Rank	Power Plant	Country	Fuel	Start of operation	Parent Company	Relative Emissions[1]	Absolute Emissions[2]
8	Turow	Poland	Lignite	1965-1971, 1998-2004	BOT GiE S.A.	1,150	13.0
11	Belchatow	Poland	Lignite	1982-1988	BOT GiE S.A.	1,090	30.1
18	Rybnik	Poland	Hard coal	1972-1978	EDF	930	8.6
19	Kozienice	Poland	Hard coal	1972-1975, 1978-1979	state owned	915	10.8

Table 1.4.: Ranking of Poland's biggest emitting power plants according to their level of efficiency

[1] Grams of CO_2 per Kilowatt hour (g CO_2/kWh). Where two plants have the same relative emissions, the plant with the higher absolute emissions (million tonnes CO_2 per year) ranks dirtier.
[2] Annual emissions for the year 2006 in million tonnes of CO_2 (mtCO_2)

Chapter 3:

**Centralized and Decentralized Power Generation-
Just in Time Solutions**

The traditional methods of power generation are under unique scrutiny for ecological and economic urgencies. Global warming is now a science and not simply "feel good" hyperbole. Depending on your non-partisan view of the earth's condition, a few general statements of our current condition may be appropriate. This we know:

1. The earth's temperature is rising.
2. The ozone layer of the atmosphere is being depleted.
3. CO_2 emissions are being captured by a gaseous blanket surrounding the earth and causing the world's oceans to become warmer.
4. In the oceans, water molecules are expanding with the increased heat, and subsequently the ocean levels are rising.
5. Coastal waters are rising at .36 inches per year.
6. Fewer, but more severe hurricanes hit the US coastal shores (7 of the strongest 10 in the past ten years.)
7. More extreme weather conditions persist as each year passes.

8. The polar ice caps are melting. Compare satellite images from the 1970's and today.

9. Droughts are more frequent and much more severe. Several nuclear power plants in France and the US have been temporarily shut down due to lower water levels and potable water needs for nuclear plant cooling.

10. Water resources are being diverted towards irrigation to satisfy agriculture needs, as well as new/old city potable water needs.

11. Fossil fuels emissions, especially coal, (50% of US generation) have an unacceptably high CO_2 emission contribution. The 2.5 Micron particulates are contributing to greenhouse gases, as well as causing poor breathing conditions for people in urban environments. Of the Dirty Thirty, number one is in Greece. Ten are located in the UK, ten in Germany and four are in Poland. The Dirty get rich. This will change ligimite coal as the predominate source of coal.

12. Agriculture products to energy products have contributed to a 100% increase in our weekly grocery bill from 150 to 300 dollars.

Nuclear power has all but run out of capacity in the US. At this time, nuclear power is responsible for 19% of the total US energy footprint (up from 5% in 1973) and 40% of the total capacity in the northeast. Nuclear power is over one million times more effective as an energy source than most fossil fuels. We have 25% of the world's nuclear capacity in our 440 facilities. Of the 103 facilities in the US, most were built during the same ten year period and are likely to be decommissioned during the same ten year period between

2020 and 2030. As of now, three are under construction. The older facilities were built with 20 year licenses in mind. The licenses were extended to thirty years, and with maintenance and improvements now have sixty year horizons. This is good news. Some experts say they can extend the useful life of a nuclear power plant to one hundred years, however these experts will not be around to validate these projections. The waste left behind from the nuclear facilities has been a "hot topic" since 1979's Three Mile Island shut down and the disastrous Chernobyl meltdown of 1986, which caused over 4,000 cancer related deaths. The benefits of nuclear energy are extraordinary, but the concerns are real. (The storage of nuclear waste, since the beginning of nuclear time is about the size of a football field and growing at 3 feet a year.) These waste containers are located from Washington State to South Carolina. When Yucca Mountain opens, securing the transportation of the waste materials will be a unique concern. Waste water for cooling on site is of concern as well. If that waste leaks into aquifers and spreads it would be bad, very bad.

Environmentalists or contrarians to the reinvestment of nuclear power argue that the markets should prevail. Why should we subsidize "Horse and Carriage" technology when automobiles are available? "Green" proponents point to the over $70 billion in investments for renewable energy, and nuclear power received nothing from the private sector.

The conversation quickly turns emotional. There is clearly "cherry picking" of data points on both sides, but the renewable energy community does so to the point of ridiculousness. That's like saying we have no need for building bridges or highways because we have aircraft of various sizes! Some emotion is driven by a 300% increase in our costs to filling our gas tanks. Groceries have also doubled. Now it costs $1 million dollars a month to $3.5 million dollars a month in electricity to power a 75,000 square foot data center in New Jersey. The cost of the subway fare went from 4 dollars to 12 dollars per visit through Staten Island all in the past 8 years.

Nuclear power was a scientific gift to the world. Today, with centralized power challenges of limited resources and 2.5 nm particulates of unique greenhouse gas concern, nuclear power is without question part of the "technology push and market pull" of energy creation solutions.

Nuclear power is mandated and mentioned by name and facility in the US Senate Bill S.2191. Tennessee Valley Authority (TVA) is the recipient of the funds to bring on line a partially built facility. Fast is good. The unfortunate facts are that nuclear power plants take thirteen to fifteen years to site, design, build, commission and operate. The facilities have unique challenges of needing to be near reliable or drought proof water for cooling and not near a dense human population for the "not in my back yard" crowd. These are two very real and time consuming challenges. It is not easy to locate a nuclear plant in today's society of "not in my back yard."

Today, most energy companies are "quarter driven" for returns and desperate for EBIDA positive statements. Remember the "dot-gone" days of silly spending on half-baked ideas vetted by the "experts" using new math to explain the ridiculous PE's and expectations of future earnings? (These energy companies both renewable and central are compelled to get a product out the door and keep expenses low.) Other than research and development, what companies make physical investments over ten to fifteen years? Even chip companies are now "fabless," meaning without facilities. They sell intellectual property and have some third world country manufacture their goods. The chairman for one large energy solutions company was "out of the closet" regarding the need to satisfy analysts and get the goods to market. Representatives in the same company are unable or unwilling to save hundreds of millions of potential dollars in energy and other costs if they spend a few strategic dollars today. This is the dysfunctional world we live in where the markets cannot and will not turn their ships.

The government's duty is to subsidize nuclear power by financing bonds, as they do for highways, bridges, reservoirs, trains, the DEP and other critical and vital parts of the US infrastructure. This is not easy stuff. Contrarians should be held accountable for such reckless statements and headline grabbing. Nuclear power is not an energy fix, but clearly part of the collective energy solution.

Demand for power in the mission critical world rises by 20% annually, and 2% for general usage. All of our conservation efforts still have us with positive absorption of energy. Coal is our greatest fossil resource, however it takes over one hundred coal filled cars to power a central plan daily. Many scientists think the development of clean coal technologies will help satisfy our short term needs and alleviate growing CO_2 emissions concerns. This solution, like most of the others is likely to give short term financial impact or an increase of cost of power and decrease efficiency. This is expected with existing and alternative methods of generating power.

Known global reserves of oil will last for approximately 40 years, natural gas for 65–70 years. Most coal reserves are found particularly in North America and China. Coal is unfortunately the dirtiest of all the fossil fuels. The two methods of energy production from coal to alleviate CO_2 emissions are:

1) Coal Gasification
2) Coal Sequestration-capture and storage of carbon

The interest in alternative fuels has been recently inspired over the past 10 years by:

1) **New technology "push"- The development of low emission technology (to be discussed later) both publicly and privately funded.**
2) **Market "pull" - Legislative incentives for low emissions compliance. Tax benefits, "green color" jobs (over 200,000 by 2020 if one fifth energy is alternative or renewable by 2020), corporate governance compliance, etc.**

The following is a summary of renewable sources of energy and installation costs:

Resource	Capital Costs Per KW
Geothermal	$2,400-$6,000
Hydro	$1,500-$9,300
Coal	$2,100-$2,800
Nuclear	$3,000-$8,500

Geothermal. Geothermal energy is created from variable heat harvested from the earth's core and the (radioactive) decay of the center. It has 50,000 times more energy than the collective fossil fuels we are aware of and produces 50 times less pollution (CO_2 and particulants) than fossil fuels according to the Union of Concerned Scientists. Currently, over 300 areas in the US use geothermal energy, but this percentage contributes to just one tenth of one percent of power globally.

The largest facility in the world is located outside San Francisco with 22 plants and almost 900 megawatts and expanding! In the US, Alaska, Arizona, Hawaii, Idaho, New Mexico, Utah, Oregon, California, Washington, Nevada, Alaska, Montana and Wyoming use geothermal energy. The Philippines has the second largest plant and are looking to build and be home to one with 1900 MW installed and reserved up to 4790 MW. Well known geothermal sources globally are in Iceland, Switzerland, Turkey, Italy, and New Zealand.

Most data points published regarding these sources are often incorrect. In Iceland, for example, 4/5 of the power comes from glacier hydro and NOT geothermal. The access to the power is limited to geographically active regions of the earth's surface.

Energy comes from heat below the earth's surface. Common sense prevailing, the deeper that one drills, the hotter the environment becomes to create steam that will power turbines. Also, common sense prevailing, this form of energy is more expensive to build and maintain. The percentage of construction costs are as follows:

8,000 ft vertical (non elliptical) well
47% Plant Construction
25% Well and Field (lay down area)
14% Exploration
5% Well rig
5% Other
4% Professional fees

The above is a model to create 5.5 cents per KWH of power.

A vast majority of US geothermal potential power is west of the Mississippi at a depth of 4 miles beneath the earth's surface.

West of Denver, it is 3.5 miles deep.

Northwest of Arizona, it is located at 2.5 miles deep.

At 1.5 miles, it is spotty in Utah, Denver, Oregon, and Washington

Power generation costs have come down in relationship to technology advances of drilling and capturing hot water and steam since the 1970's. The cost of geothermal power is between 3 and 6 cents per KWH, a remarkably low cost for a reliable and consistent source of energy. The sourcing of the energy and construction is time consuming, but far less time sensitive than nuclear facilities and dams. Saline fluids are a maintenance concern for blockage, but this concern is manageable. Geothermal energy has legislative support and it should in order to keep it visible. The US is rich in the resource, predominantly out West. The tax credit established in 2005 was set to expire in 2008, but a bill is currently pending to extend the credit to 2012 to support development. A unique interest in oil rich parts of the US is to reuse dry oil wells and sink them deeper to hit heat for steam creation. Due to the increase price of steel and labor, drill costs have reached $2,700 per meter and continue to rise...it still makes sense with the rising costs of fossil fuel and carbon credit benefits in place and forthcoming. Iceland continues to make global contributions for commercial and residential geothermal energy creation and domestic space heating, potable water and sewage. A deep drilling project at Hellistheidi will be 4-5 KM deep tapping 400-600 C subsurface hydrous fluids to manipulate, if successful.

Hydro-Power. This form of energy is already a mature technology. It accounts for about 20% of the world's total power generation. Under appropriate geological and weather conditions, hydro power is a very compelling source of energy. Larger or "mega plants" are some of the

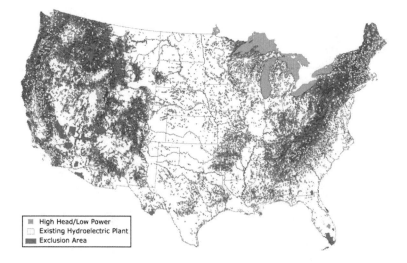

High Head/Low Power
Existing Hydroelectric Plant
Exclusion Area

lowest cost electricity generating sources in 2008. In the United States the top 5 hydro producing states are Washington, California, Oregon, Alabama, and New York, in that order. The 1-2% annual growth rate is well below the renewables like "other" (wind-20%). China, though, is providing an increase of 160 to 300 Giga watts by the year 2020.

China is the number one in hydro power production with the US, Canada, Brazil and Russia rounding out the top five producers. Many of these plants were built long ago with sunken or fixed costs which have long since been retired. The methane emissions from plant and animal decomposition are twice as effective as CO_2 for retaining heat or greenhouse gases. This is an issue for all the dams, especially

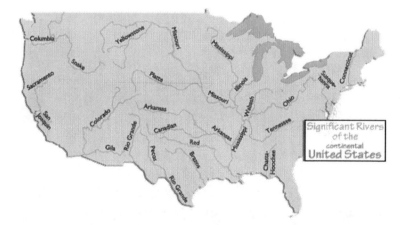

those with shallow reservoirs. Hydro plants contribute approximately 10% of our nation's power and 20% of the world's power. At 3 to 4 cents per kilo-watt hour, it is a compelling energy source, however many of the appropriately placed dams have been built for centralized energy creation. Micro dams, or smaller dams will continue to be built but there are longer permitting and construction concerns. Another roadblock to building a hydro-power plant can be the ecological considerations and sensitivities which are scrutinized carefully for ecosystem issues like fish spawning, aggregating silt, methane distribution, etc.

As for wave technology, is correctly deployed for about 2,000 Giga watts or about an eighth of the total global energy used. The tidal power is promising and deployed. The largest facility is in France at 240 megawatts. South Korea is to finish construction of a 254 megawatt facility by 2009, and an 813 facility is planned by 2013.

Chapter 3
Centralized and Decentralized Power Generation-Just in Time Solutions

Wind Power. The power of the wind provides a minority of alternative energy at approximately 1% globally (74 Giga watts). This is expected to double by 2010. However, it is 6% of Germany's, 8% of Spain and 23% of Denmark's total power usage. Worldwide, wind energy has increased 400% from 2000 to 2006 with over 100,000 wind turbines in over seventy countries. Floating deep water wind turbines are likely solutions to energy creation with blades over 260 feet in diameter. Two thirds of the planet is water and wind flows freely over water. 2.3 megawatts wind turbines are built in concept from oil rigs floating technology. An $80 Million dollar Norwegian project is expected to come online in 2009. The floating draft will be 1,000 feet below the surface and anchored to the ocean floor. Very smart.

Wind energy represented approximately $18 billion and was expected to reach $80 billion by the end of 2007. Texas has the country's largest wind farm in the US with North Carolina building a larger one at this time. Do not get too excited. Energy Storage is still an issue and the cost to tie into the Grid is a multi-billion dollar challenge. The backlog for wind blades from large manufacturers is currently 3 years, and that is only for large orders. The costs to produce wind power varies significantly based on field conditions of land or sea The American Wind Energy Association (AWEA) indicates the 1,400 Mw or 3 billion in wind energy was produced Q1, 2008 which brings the total US installation of 18,000 Mw with 4,000 in production. The cost differences are as dramatic as 5 to 25 cents per kilowatt hour. The permitting and development requirements have long durations. Factory orders are full for the next 3 to 4 years. The nature of the wind volatility and speeds is a concern for stability and consistency to push these blades, which are as long as a football field from end to end, turn 15 rounds per minute and weigh 4-6 tons. They do kill approximately 10 birds a year, which may be considered the culling of the flock since the blades are 6-8 feet wide and are about 140 feet long. It's not like you can't see the 15 RPM blade moving!! These would be the birds on the "short bus" of life. The source has captured the confidence and imagination of the public with wind farms growing by over 20% since 1990. This is encouraging for the future use of wind power, both centralized and decentralized.

Solar Energy. This form of energy is widely used commercially for transmission and distribution. The largest plant in the world will be built outside of Phoenix as a 280 megawatt solar/thermal plant. A 64 megawatt facility in Nevada has 180,000 mirrors on over 400 acres of land that gets sun 330 of 365 days a year. You need a lot of sun and a lot of real estate to site those panels. Japan has deployed approximately 40% of the world's solar power.

In the US, investments from the private sector rose 15% to over $450 million dollars. Solar stocks have risen 300-3,000 percent and are at this time being "shorted" due to excessive exuberance in this source of energy.

The growth for environmentally aware public and commercial deployments of 50 megawatts and greater in California are successful and ongoing. A dirty little secret of the panels (PV) are that they create HAZMATS as waste products in both the manufacturing and future disposal. An extraordinary amount of energy is used to make the panels, including the lead based or lithium batteries that store energy during darkness and foul weather. At 10 to 20 cents per kilo-watt hour, this is not a cost efficient alternative for many parts of the world. Solar energy is a market ready technology for residential and smaller non critical load demands. More recently and due largely to improvements in better technology and energy storage, large scale commercial deployments nationally and internationally are in use and growing swiftly. Battery or energy storage is a challenge, but salt in insulated tanks show some promise. Converting AC transmission technology to DC to help long distance solutions for (Midwest or long distance) generation for DC to urban environments is now considered viable.

Bio Mass. This is not a zero emitting technology. Through photosynthesis, plants produce sugar from the atmospheric carbon dioxide and use sunlight and oxygen as a by product. The process is reversed when bio fuel is burned. (Bio-waste captures gases-methane to create energy and turn the turbines). The energy that has been stored is then released, coupled with the release of carbon dioxide. Invested dollars in bio fuels in 2006 were over $800 million.

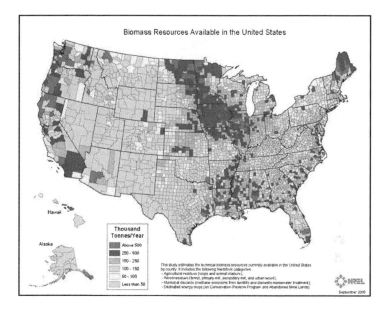

The DOE is trying to get the country to double the 1% of electricity and 2% of liquid fuels currently in use by 2010 and 20% by 2030. It is not likely this will be achievable with today's development and distribution, specifically distribution. Where to buy the product? Municipal or fleet cars and trucks for mass transit have a better chance of making short term contributions with fixed fuel stations and service station depots to support a fleet with predictable routes and fuel storage/distribution. To turn a fleet of cars and service stations via an "onesy twosy" marketing plan would take 8-12 years for people like me to buy a car and wait for the service and fuel stations to pop up on likely and frequent travel routes. Collectively, it is a bad bet at this time. When used in burning for transportation (cars and buses) a 30% to 90% savings or reduction in carbon emissions and 2.5 micron PM (particulant material) is achieved when compared to gasoline. Net savings are realized with blended solutions proportionate to the bio and fossil components. Costs range from 4-7 cents per kilowatt hour, depending largely on the size, scale or distance of the biomass.

Transportation is another sunken or "well to wheel" cost often overlooked in alternative energies. Bio fuels can be produced from a number of crops and plants; corn, canola, palm oil, sugar cane, and jatropha are the top five. The following are tons of bio mass currently in production:

Corn-750 million tons...primarily in the US
Canola-46 million tons....primarily in EU, Canada, China and India
Sugar Cane-3 million tons....primarily in Brazil
Palm Oil-35 million tons....primarily in Indonesia and Malaysia
Jatropha-unknown....thousands of hectares panted and harvested in Africa, India and the Philippines.

A World Bank study and an investment anti poverty group reported that there has been a 30% increase in global food prices due to this shift in energy crops.

Bio Waste. It is not a new idea for energy creation. Anyone who has driven by a municipal waste dump at night has seen the yellow and blue flames of spontaneous combustion caused by methane. 65-70% of our waste is degradable and will create gas. In New York City, we flush 80,000,000 gallons of waste daily. That sounds about right….8 million people…5 gallons per flush and two flushes per day.

The solid waste from these flushes goes to the 14 waste stations at the water's edge for barge transportation and is sometimes used as energy creation via the capture and burning of methane or the drying and transporting of waste for fertilization.

In the 16th and 17th century, manure was commonly transported by ship when necessary and by land in smaller containers. Wet manure is heavy by nature and large shipments on wood ships were the most practical solutions of the day. Merchants would dry the manure prior to shipment for ease of movement. When manure is stored at sea, for longer distances, the manure would ferment when it eventually got wet from rain and moisture. The methane by product was recreated with decomposition and the moisture or water-and you guessed it…….an explosion! When a ship mate would bring a lantern (fire) below to have a smoke….BOOM!

Several ships were destroyed and many more damaged before the cause of these explosions was fully diagnosed and made preventable. After the diagnosis, the manure was stored in the open air and above the main deck with the stamp "Store High in Transit" or SHIT. This is comparable to our "This End Up" or "Fragile" notices we put on cargo bins today. Bio waste energy creation is an outstanding use of animal and human waste. The carbohydrate energy values are about 35% of fossil fuels, and the carbon footprint contributions are minimal. Human waste or cow manure methane capture is a standing and effective technology. Methane from a cow (let alone the 300 million humans in the US) can generate 100 pounds of manure a day. That's right! A cow excretes 100 pounds a day. Through digestion and decomposition, this makes bio gas and mostly methane with the addition of oxygen. Human waste and animal waste creates methane to burn and drive generators stored from restaurant's produce or waste oil.

This is well done in France with their buses currently. The "Lille Metropolis" community represents an area of 85 cities and 1.1 million people. 127 of 311 of their fleet of buses run on a combination of natural gas and bio gas. The bio gas is created in bio-waste facilities from kitchen and garden, municipal and food remains. It takes some 34,000 tons of the compost to produce 3 million metric tons of gas. The bio gas contains 65% of methane gas.

This concept works well when the goals are clear and the use, fuel, creation are supported by the public or government. Similar mass transit and, bio gas programs are in various stages of design and development in Stockholm, Prague, Graz, and parts of Europe.

In the US, landfill gas projects are not new but are attracting new attention. We now have 450 operational plants as part of the "Landfill Methane Outreach Program" (LMOP) with approximately 540 more planned. In other words, more highly populated states have more waste, and therefore more landfill with methane opportunities. The top 5 methane states via landfill according to environmental petroleum agency midway through 2008 are:

State	Operational Plants	Candidate Landfill
California	72	37
Illinois	35	24
Michigan	28	9
Pennsylvania	27	15
Wisconsin	22	10

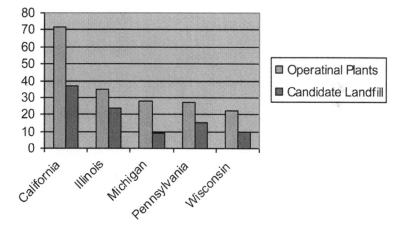

Ironically, New York and New Jersey have 19 and 17 respectively. The states obviously have tremendous waste, but are exporting it to other states. Texas, Florida and South Carolina are large recipients since Fresh Kills was closed on Staten Island. The waste is transported by land and sea in many cases. This will likely change since both states import a tremendous amount of energy to their large cities.

Ocean Energy. Often called "Blue" energy, ocean energy is wave or tidal energy. Wave technology captures the vertical inertia movement consistent with the movement of a buoy. The rhythmic up and down movement creates energy which is transmitted to land, where it is synched to distribution conductors.

A company called "Wave-Bob" in Europe has deployed enough power for 1,000 homes in Galway, Ireland. Wave technology captures the tidal movements in a dam or restrictive way to move the paddles of a breach to turn a turbine in synch with a generator. Water is also captured and released on both sides of a tideflow to keep continuity of water flow over the paddles, minimizing the hydro benefits of man-made dams. This technology has successfully been used in Europe for hundreds of years and in the US for the last 40 years. A range of 8-12 foot tidal movement is the most effective. This is a brilliant alternative for appropriate regions of the world. Turbine power from predictable currents and rivers is exceptional as well. The water flows through cylinders forcing water to move turbines to start generators and cable energy to land. A bit disruptive to marine life, but a huge energy win! Image water front or river prosperities and business buying their energy source at Lowe's or Home Depot. It's not out of the question.

As users, and particularly for large users, we ask ourselves a fundamental and philosophical strategic question every 3 to 7 and sometimes 12 years.

The 3 to 7 to 12 years are milestones of time which prompt the user to make strategic decisions on real estate, operating expenses, growth (organic or acquisition) size and location, etc. The globally recognized lease terms for most transactions are 5, 10 or sometimes 15 years. Where do I move my offices? Do I renew or move? How can I impact my operating expenses? Trust me, utility costs and reliability issues are "in" there! Yes, there are shorter term leases, but there are longer term leases which have breaks at earlier time frames.

Lease terms beyond 5 years are really just best bets based on relevant information available at the time of the Steering Committee's efforts. This information, of course, can be totally sidestepped or ignored. Another factor can be a chairperson's heavy handed interest in working at a location close to his/her home, state or with a lease duration co-terminous with their employment contract.

Having said this, a disproportionate amount of time is currently being spent during the strategic phase of siting industrial, manufacturing or office plans based on the operating expenses of a location, city, state or region. The drivers for operating expenses are:

- Power
- Real Estate Taxes
- Personal Property Taxes
- Sales Taxes
- Human resources (churn)
- Weather (lost time/quality of life/inefficiencies)

What we have recognized over the past ten to fifteen years, but more so recently, is the extraordinary effort businesses are putting into operating efficiencies. It's like "found money" impacting in many ways the bottom line significantly and translating into a more favorable PE in many cases. These efforts are called "treasure hunts" for energy conservation companies.

The auto industry uses more computers or robots to build cars, thereby taking them away from union discussions or health care commitments. Government and technology companies are using more "flex" time or work from home options reducing the real estate commitments and associated operating expenses. Fewer toll takers are on the bridges, tunnels and toll roads. Chips now electronically collect our fees to pass. Buildings turn off lights automatically and remotely; they monitor the sun's location and associated cooling needs. Elevators in office buildings "rest" in strategic locations ready for "time of day" regular movement for the next lifts. Smart buildings are designed from the inside out (technology based) rather than the outside in (cosmetically based). We have migrated slowly from the post modern marble and millwork-clad lobbies of assets meant to lease and not to last, to forward thinking power and water efficient assets better prepared to endure resource scarcity and unplanned outages. Co-generation and hydrogen fuel cells supplement total energy loads and contribute to meaningful LEED credits for new building and retrofits. Federal, State and Local incentives are required in the short term to make the business arguments compelling in the 3-5 year modeling and slow the demand for centralized power creation with traditional costly and non-carbon friendly technologies. This is what government does. It shows the way. It shines the light on best practices and often provides the technology "push" and financial "pull" required to help the free markets help themselves.

We are making strategic decisions based on three main drivers that will have an inescapable impact on the economic and lifestyle goals of a user(s):

1. Power-quality and availability.
2. Resources-water, air and other.
3. Weather-history of adverse events and favorable lifestyle changes.

The relevance of centralized vs. decentralized power is a main issue for the long term strategic planning and growth. The centralized power plant of hydro, fossil and nuclear power were a good offensive and defensive play or strategy the government made given our natural resources, growing demands, and diversity of the earth's surfaces/ conditions throughout the United States.

Military specifications have been the historic cornerstone of some of our best thinking and planning as a country. It has been the home of the country's best and the brightest. As it relates to power generation and transmission/ distribution, their philosophy of "two is one and one is none" is their mantra. They live this philosophy whether they are packing socks for a trip or designing a nuclear facility because shit... happens.

Our regulated power system of twenty years ago and up until ten years ago was generated by almost equal parts of nuclear, hydro and fossil fuels, with a 20%-25% plus reserve capacity for a planned or unplanned outage. Our utilities had reserves to handle generation and transmission challenges created by seasonal challenges of events that of any one source could provide a seamless transfer of power in event unbeknownst to the user.

Centralized power was the result of super sized energy creation plants of various sources built to last multiple lifetimes. These went but not always in the most strategic locations. Super sized dams, coal plants and nuclear facilities all had large and redundant capacity plans baked into the strategic vision of the asset. We, as tax payers, expected nothing less. Our urban environments were fed power on hot or cold days without interruption in most cases. They are massive importers of energy. New York City is responsible for one third of New York State's energy demand, and when Long Island is included, they account for one half the state's total energy demand. The networks were available 99.8 percent of the time. That 2 tenths of a percent translates into 4 hours per year cumulatively. In a non-mission critical world, these interruptions of service or distribution were acceptable, and by the way...there was no recourse for interruption. Unless there was some form of "willful misconduct" or recklessness, there are rare consequences for poor planning or implementation of energy creation, transmission or distribution.

The centralized power models post deregulation has far less reliability and reserves currently. Twenty percent plus reserves are down to 12-15% in 2008. As discussed earlier, nobody ever got rich with high inventories in any business. We do not reward computer chip makers for high inventories after the holiday season or auto makers for high inventories at year end.

A decentralized model of power is the current answer to the legacy question of: how do I control my own destiny of power as an operating expense? How do I influence the reliability and redundancy? How do I future proof this critical component on my operation that we take for granted? How do I depart from the "lemmings" model of they "bill" I "pay" and pass on increases to my customer model that is making the US less competitive? How many of us know how to read or interpret our utility bill? Users need to ask themselves some strategic and tactical questions on how to meet today's energy challenges. They need to understand their usage, methods of billing, tariffs and incentives (public and private) and get as granular as reasonable to understand and control their energy usage.

Decentralization of power is the concept of creating power on or close to one's operations. This works for governments, schools, manufacturing, pharmaceuticals, mission critical data centers, military, industrial and cluster residential applications. Just think of the time, expense and maintenance it takes to transmit power (regardless of source) across cities and states to one's property line. It makes practical and financial sense to keep power sources close to home base.

If the power was gas for co-gen, geothermal, biomass, solar, wind or other, the creation, and distribution would be in smaller or scaled units for better overall accountability, scalability and maintainability.

Decentralized power takes the large operating expense that has gone up by over 100% over the past five years in many parts of the country and puts the one time and future expenses in the user's hands. This enforces better efficiency and a more favorable carbon footprint which will likely earn credits and payments monthly from the incumbent utility to stay off their grid and make power available to the grid when needed.

Decentralized power, not always but often has an eighteen month to eight year payback on the return of the investment. For co-generation (or CHP) the cost of gas and existing price per KWH are the big drivers to the business model. It gets complicated when the price of gas has hit or met the utility market due to long term contracts maturing and rising. The price lags for months, but rarely years for gas. It costs between 1-3 cents per KWH to operate these facilities in regions that have tariffs of 1-4 cents per KWH. This makes "cents" and is good business.

The quick math is that they can often pay for themselves with commercially deployed proven technologies within three to five years and operate at a profit of almost 100% in a carbon friendly model. The three components of co-gen that need to work in favor of decentralized co-generation are capital expenses, cost of gas and maintenance.

So why do we not see more decentralized models or power plants around the country?

1. **The utilities make the permitting and construction requirements protracted nightmares in most cases.**
2. **In low cost power regions of 4-8 cents per KWH, it is not economically compelling in the short term.**

3. **The government has not cut the red tape or led the way to sustainable alternatives for long term solutions with short term permitting and construction urgency.**

4. **Not enough commercial choices of quality vendors to choose from who will be in business for maintainability certainty. Many of these companies do not operate in an EBIDA positive model currently.**

The centralized power model worked well to serve dense urban environments. As the cities grew larger and out and around the power plants and substations, modular growth become unrealistic and that's why we buy and sell power out of our region in most cases.

Suburban sprawl has caught up with remote power plants where residents don't like their kids playing near substations, HV or LV power lines. People also do not want to be near nuclear power facilities or water near nuclear power facilities. Residents want to be as far as possible from gas or coal fired plants which cause asthma and other health problems. By the way, New Jersey is the asthma capital of the U.S.

Smaller, cleaner, and more nimble power solutions make more power sense to meet supply and demand market conditions as well as reconciling operating expenses, reliability, and expandability and maintainability goals. These strategic "guerilla" warfare solutions are being tactically implemented successfully.

Chapter 4:

Post Deregulation Realities… Follow the Money

Under the former regulated power structure in the US, like telecommunications and air travel, the regulated energy system was a monopoly. The utility was assigned a geographic footprint to create, transmit and distribute energy. The power company had the exclusive rights to create, market and distribute power to the end user or customers. Quite simply, they were franchises. The fact that the utilities are no longer regulated should not be indicative that markets for electricity are un-tethered or free with choice. Not a play on words, rather users or tenants in certain regions or "footprints" have no "choice." The spirit of freedom of choice via deregulation does not really exist in most parts of the US. Tarriffed footprints are regulated and managed under penalty of law for the seller and buyer and the "last mile" incumbent gas distributors cost is protected by law as well. This is not negotiable.

Before 2002, the franchises would make assumptions on existing needs and apply a complex polynomial equation (lots of numbers) for growth. They would assume changes of new needs or losses of loads based on the regional economics and build to suit them. They would then apply a relevant maintenance structure around the fixed assets inclusive of the network operating control centers, (NOCC's) including house repair and replacement equipment kit strategically for fast and critical service (rolls and poles) and apply a reasonable and defensible mark up for profit.

At the time of the "blackout" of August 14th 2003, that impacted eight states and numerous cities including New York City, compliance with reliability standards post deregulation was entirely voluntary. The Energy Policy Act of 2003 made new standards mandatory and led to the creation of the National Electric Regulation Committee (NERC) of 2006, which created and expanded a set of standards in an attempt to make our reliable "just in time" network more dependable.

To summarize: the utility deregulation of 2002, the regional multi-state international rolling black out of 2003, and the Energy Policy Act of 2005 all combined to create reliability standards. One can see the wisdom of this idea, but the devil is in the details. There are no criteria for maintenance, tree trimming, redundancy, testing or pre deregulation reserves of energy capacity. Later on, in New York in 2007, further attempts were made to enhance reliability by creating the New York State Public Service Commission (PSC). The PSC would begin to address the critical issues of infrastructure design and maintenance to enhance long term success of the network. Similar regional solutions were being implemented to supplement the design and implemented to supplement the design and implementation needs left in the wake of energy deregulation.

Prior to the regulated companies designing or building new generating or transmission facilities, they needed to get the prior approval of a State Regulatory Commission or the Public Utility Commission (PUC) in all cases. The PUC did the cost analysis of a solution for the proposed generating source expansion or new source, as well as the most cost effective transmission route and construction estimate. If approved, the improvement costs were approved with interest and carrying cost to the utility rate to the customer base. New York is barely holding onto 15% of reserves to protect the hot summer days and negative cascading potentials of the network. Currently, the same reserve or capacity model applies to telecommunications. When NYC needs emergency power due to hot summer days or other unforeseen shortages, we buy power from the very expensive stand-by generating companies or regions. The use and high costs are then passed on to the consumer.

Overall, in the regulated environment, the cost to build and maintain reserve capacity at the generating level was .5 cents per KWH. Obviously, this is a very small part of a 24 cent per KWH market, but a more impactful part of a 4-5 cent per KWH market. Most generating and transmission companies are more supportive of an organized and contractual load shedding program in the form of "back stop" or demand supply programs. The PSC believes that there is more than 25% potential in energy improvement from antiquated power plants. For instance, according to the ISO of New York State, 68% of the power to support NYC and Long Island was built before 1980. These can not be new plants, and 72% of these plants are gas or oil driven. Of new plant upgrades statewide, 9,000 megawatts are gas/oil, almost 7,000 megawatts are wind, 3,000 megawatts coal, 300 megawatts nuclear and 3,000 megawatts hydro. Of these upgrade options, coal is the fastest, but least eco- friendly. The 2% annual growth that is projected by the ISO does not incorporate the reduction of Indian Point Reactors 2 and 3, which were denied a 20 year license extension and are scheduled to close in 2013 or 2015. Plant number 1 was decommissioned in 1974. The plants in New York and other plants to close or be decommissioned are the Charles Poletti plant in Queens, Lovett 5 and Russell Station for a total loss of 1,300 megawatts. New plants are coming on line, but conservation measures coupled with organic and seasonal (anomolic) growth keeps this network at risk.

Demand is outpacing supply and conservation and load shedding policies will not alleviate the needs presented today to satisfy the 2/3 statewide 38,917 megawatt demand in the southeast and ½ of the New York State's 10,019 megawatts demand dedicated to NYC, exclusive of Long Island.

In the regulated environment or model for any industry, we assume the PUC demanded that the generating system and transmission system to be severely scrutinized, controlled and maintained to levels on or close to military specifications. It would have to be. The lifestyles, commerce, transportation and national security of the most powerful country in the world were at stake! Not only were the real consequences of poor design, implementation, and maintenance at stake in this extraordinary inter-state "just in time" network, but the perceived or anticipated failures were also at stake. It had to look and be bullet proof.

Like any widely used utility such as water or sewer, we take them for granted. We wake up as children and assume that water will come out of the faucet, the toilet will flush and the lights will go on, not necessarily in that order. If America has a brand (and it has many), one may be that we are a technologically advanced country. We can meet or exceed the challenges of the day, no matter how difficult, protracted or expensive. In our regulated world, the American brand for reliability, scalability and maintainability was satisfied. If anything, there was the image that it was a sleepy, stodgy "belts and braces" over-built, over-spent infrastructure that was too big to meet the unique empirical challenges of generation and distribution for the technology revolution.

The decentralization of the urban environments, exporting of manufacturing jobs, the migration of industrial work to points south-out of the "rust belt," coupled with Moore's Law and with the information technology (IT) related loads of voice, video and data on our offices, home offices and personal digital accessories (PDA's) have made the strategic planning and "just in time" implementation of energy daunting. One could have made the argument that due to the dynamic changes mentioned above that the 10 regions or consortiums of power generation and multiple transmission or billing entities are more appropriate.

The philosophy of the deregulation was partially based on the perception that without competition and price pressures created by the market system, there would be open to be cost efficient either in generation, transmission or distribution of power. There was a perception, that the super structure of energy was dysfunctional by size and littered with inefficiencies of scale, infrastructure, human infrastructure and intellectual capital. One could argue that there was no motivation with lifetime employment models to think outside the box in order to meet the fluid and changing landscape of power. Why take the risk? Why make changes or suggestions that may end up being career killers? Unfortunately, the "not on my watch" mentality may have taken hold in parts of the regulated community. Surely, this is part of the reason that we find ourselves in this current situation.

The market forces under deregulation sold very well at the board room level. On paper, or on the "white board," a strong argument could be made for the benefits of deregulation.

However, one does not have to look far to find examples against deregulation based on the airlines models and the telecommunications model. The economics of deregulation are compelling. This is a cost model. However, the quality and strategic planning components are NOT compelling. As well, one does not need to look too far to find a former worker of the regulated airlines maintenance business vowing never to fly under today's conditions. Maintenance and repair issues here and abroad are often eclipsed by the need and usage of the aircraft for profitability. Pilots are flying too many hours on equipment that is being pushed beyond a reasonable maintainability and life cycle window. The passenger is now subject to horrific flight delays due to the fact that the airlines often over sell the seats to ensure maximum profitability by a minimum of 10%. Then, we all have to wait patiently (without voting rights) as the "flight attendant" articulates that there are too many passengers for the flight and start auctioning off seats on later flights or unlimited travel to those willing to play airport "lotto."

In telecommunication terms, the sizing of the switch was far more liberal under the regulated model. The switches were sized in anticipation for every line to be used at the same time.

Like the airlines, they assume a certain percentage of people would use the network at the same time, at certain times of day, days of the year and so on. However, post deregulation, the land line congestion follows a large media event like a storm, ball game, elections are well known. In the Centrex world of circuits, the switches were sized and beyond that, calls were dropped. The packet or internet protocols have relieved most of those challenges. Currently, Centrex calls still do not go through because there is no profit in excess capacity or a dead asset for most of the year in the supply/demand world of free markets.

Maintenance in the air, telecommunications and energy markets have been a unique concern for the consumer and the provider. By way of example, in the nuclear energy world, during a 2002 scheduled refueling outage at the "Davis Besse" nuclear facility in Ohio, workers discovered that boric acid had cut a hole the size of a pineapple in a 6 inch thick piece of steel. If the hole had gone 1/3 of an inch deeper, there would have been radiation exposure. This is frightening to consider.

Following an investigation, The Nuclear Regulatory Committee (NRC) held the engineer in charge of maintenance responsible and barred the individual from working in that industry again. The engineer was indicted on five counts, including lying to government officials. The man claims he made management aware of the need for maintenance, but to shut down the plant for repairs would cost approximately $1 million per day. It was put off. The happy reality is that nuclear power is remarkably reliable and dependable. The consequences for the unlikely are severe within the immediate "kill zone" around a facility. The zone includes eight to ten miles, and extends to fifty miles for agriculture and livestock.

In the US, and within the deregulated market, generation is effectively the only real component of choice or product with competition. Today, given licensing and a protracted legal process, almost anyone can build or buy a power generating facility or become an Independent Power Provider (IPP). For generating companies, electricity is a commodity. The "special sauce" for effectiveness or profitability is the technology and resources applied to the region to adapt and address the weather conditions in the footprint and the cost benefit of the natural resources to create energy. The newest component on the path to profitability is the ecological considerations of emissions created by the generating source.

The generating companies under deregulation often do not have the luxury of time to satisfy the near term power creation needs. The thirteen to fifteen year horizon to build a nuclear facility, the five to six years to build a geothermal facility or the four to seven years to build a large micro-dam is an eternity for these generating companies in the "just in time" business of power creation. Remember, we took the

1990's off for nuclear power creation due to environmental concerns. Now, global warming is overcoming the unlikely catastrophic event at a nuclear facility.

Generating companies need to collectively put their money to work to yield the greatest return on investment. It takes approximately 5 billion dollars to build a nuclear facility and 10 billion dollars to decommission a nuclear facility. It also takes almost 15 years to build it....this is an extremely hard board room sell. Fossil fuels are faster and cheaper to make, but catastrophic to the environment. The slowly moving Carbon Credit and penalty programs in the US will be a real driver to get the attention of the corporate user and generating companies on the bottom line cost and efficacies of power. They have already been successfully incorporated in Europe.

By way of example, the Chicago Carbon Exchange is a voluntary exchange where a carbon credit is worth 3-6 dollars a ton per years. Mind you, the emphasis is on voluntary! The European Carbon Exchange is now worth 25 Euros or 37 dollars per ton and is mandatory! The Regional Greenhouse Gas Institute is looking to set the market in 2009 with mandatory and enforceable limits of carbon creation and credits for offsets of carbon creation. There are millions of dollars a year in added savings, depending on which side of the transaction you are on for annual consideration. See the Congressional Budget Office (CBO) estimates for the National Energy Security Act S.2191 of 2007 for the per ton values. They publish a value of $24 per ton, but they currently trade in Chicago on the CCX at 2-4 dollars and sometimes

6 dollars per ton. The $1.9 trillion in revenue (developed by the congressional office of the budget-COB) is based on numbers we have not even gotten close to in the to in the U.S., (5 dollars per ton vs. 24 dollars per ton) The Kyoto Protocol does not even consider intercontinental credits or trading as we speak. Look for Kyoto Protocol 2 before 2012. Kyoto has galvanized the world for a worthy cause, but falls short in content in many areas. Similar to the War on Drugs or the War on Terror, a War on Greenhouse gasses would not be impactful.

In the deregulated market place, currently the wire companies of transmission and distribution still hold monopolies within their respective footprints. It is fairly uncommon to find choice within a particular region. From time to time, if you are a big enough user and can make distribution a viable and profitable scenario for a coop...there is choice. These are one off opportunities and do not exist in any equidistant or strategic locations around the country. For those IPP's or users trying to sell to any excess capacity, (or reserves from self generation) the financial and litigious challenges are both real and protracted.

As energy users, if we want to get large capacity from a distribution company, the first thing the distribution company wants is a check...money. They are looking for between 10 and 30 thousand dollars to pay for in-house engineering time and deliverables to articulate the new demand. Realize that salaries are sunken costs to maintain the network. This is called "double dipping" in some parts of the US. Double dipping is a condition where the rate payers are paying the salaries and so are the users in search of solutions to unique power concerns in the form of consulting fees. It is also good business.

In their defense, the transmission companies, were over-used and one could say abused by the over-zealous –over-night-experts born from telecom deregulation looking to site and build these high density assets in high demand. Any building with a high ceiling and heavy floor load was a good candidate, and ALL the real estate brokers were working with credit worthy and REAL users..... not. Unfortunately, more than half the demand from Competitive Local Exchange Carriers (CLEC's), Regional Bell Operating Companies (RBOC's) and Long Distance Carriers (LD's) are out of business or consolidated.

Sadly, plenty consultants in this space are mentally challenged as it relates to "inside" plant and "outside" plant power challenges and realities but that did not stop them from taking valuable time and resources from the utilities. There was not enough respect for the engineer's time or knowledge by the brokers or underwhelming consultants. These people (network engineers) are incredibly important to our network and the real backbone of part of the US economy.

Fast forward...for 90% of these requirements, most loads never show up or those that do never exceed 50 watts per square foot from a designed 100-150 watts per foot. The funds they request and are currently requiring turned out to be a cottage industry to pay salaried personnel to effectively do the job they were hired to do. More realistically they want to sort out the real requirements from the waste of time efforts or "go sees" from knucklehead brokers or consultants.

The challenge of describing or guiding the user through the transmission piece of deregulation is that the rules change from state to state and also from utility to utility. Each state has created or defined loosely the rules of engagement for cost per KWH, taxes for same, reservation fees for day one and future loads. The capital expense for substation improvements, controls, monitoring, service, conducting, easements and ROW work, for secondary feeds and for alternative/future loads are all but "rough justice" and are merely approximate pricing models. At the end of the day, it is the real cost plus what the market can bear if there is inter or intra state competition. In reality, the user has few or no choices in a fixed or tariffed footprint. The choice is really to move or "vote with their feet" as they say.

Sometimes a user tries to self help their operations by creating their own energy source with or as an Independent Power Provider (IPP). Tying into the network for selling excess capacity or using the utility as a redundant back stop for planned and unplanned outages for maintenance or failure is often a litigious nightmare.

We collectively embraced deregulation and the free markets to guide us to a cost effective way to create and distribute a more efficient, resilient and reliable method of creating and selling power to people who have absolutely no experience in the free market power paradigm. Worse, is the model that the executive level of strategic leadership went to school on "real time." The first financial cuts on the path to profitability is early retirement or reductions in force (RIF's). The companies are losing their greater assets....their people. Only worse than that is the voluntary migration away from the model by the best and the brightest employees. They generally leave first and with them the intellectual capital that made the companies great. Effectively, that leaves the "B" team to execute the "just in time" network which most of them have never done before.

The yardstick of success is measured by profitability in the deregulated world of energy. It is the wrong metric in both generation and transmission/distribution. The markets and market value are driven by the return on equity, return on investment, cash flow, dividends, PE ratios, etc. The metric for successful generating companies should incorporate the reserve ratios maintained to protect our national security and economy with alternative fuels and progressively shoring up our traditional and expiring methods of generation.

Generating companies need to incorporate an increased velocity of adverse meteorological conditions like rising coastline water levels, drought, severe hurricanes, interrupted transportation and make sure the "just in time" utility machine that makes the economy and lifestyle function to make it's future proofed. The increased velocity refers to the 10 largest "named" storms which occurred between 2006 and 2007.

Transmission companies need to make sure the distribution and substation network that is aging is scalable and burstable that can prevent negative cascading. Granted, building the entire network at 100% redundancy may be cost inefficient nationally. However, better monitoring remotely of spikes and failures with automatic transfer switches (ATS's) which work without humans rather than manual throw over's (MTO's) which require human intervention in the field, would enhance our network considerably.

The cost of power is not a negative consequence of regulated or deregulated power creation or distribution. It should be expected. The cost of copper has gone up by over 100% in three years. The cost of coal has gone up by 300 to 400% over the past 4 years. The cost of gas has gone up by over 40%, and the cost of uranium has gone up by 60% in the past two years. The price of labor has gone up 15% since 2003 in most market places, and the overall cost of general construction in the nation has gone up by 31% since 2003. Costs have risen even higher than that in work that is weighted heavily with electrical and mechanical trades and equipment. So, how is the price of power to go down without some severe compromise to the integrity of longevity to the infrastructure let alone work along with the Kyoto Protocol, Clean Air Act of 2002 or Clear Skies Amendment of 2005?

Given that CO_2 sequestration can cost a legacy fossil plant 10–15% of inefficiency, and that generating companies, are not rewarded for large inventories or excess capacity, there is a "perfect storm" of negative supply influence and growing carbon friendly energy creation demand forces that are not in synch. Inventory is a dead asset and not profitable. Companies with low inventories are financially rewarded for running a "lean" organization. These are inherent corporate qualities we do not wish for our deregulated companies unless the reserves can be drawn upon with little damage to the environment like hydro, geo thermal, solar and tidal. Assets like nuclear are so expensive and take so long to build, the cost of capital and interest make them hard to sell at the board room level without government bonds or subsidies. They are almost impossible to finance.

Transmission companies are not rewarded for changing out legacy substation transformers or controls requiring human intervention rather than simple automation. These conductors that have been hanging in the elements for 40-60 years and badly need repair and replacement. The less than stellar general maintenance programs cut during deregulation have taken their toll in circuit reliability. Drunk drivers damaging poles and repair instead of replacement too often are the financial solution. Poles are now shared by the telecom companies and by the utilities make's ROW's more vulnerable as a single point of catastrophic failure with pole specification for telecom and utility facilities being dramatically different. Today, where substations need to physically grow due to suburban sprawl or new needs, they are landlocked and can not grow. The alternatives to supply "the need" by the transmission company is to site or reconductor a "new" route which is often not an efficient Euclidean path or point to point, but rather a circuitous path, avoiding the "not in my back yard"(NIMBY) crowd, and given to the low bidder. This does not give the warm and fuzzy feeling or cost containment or quality of power, does it?

The point here is that for transmission and distribution companies to ensure reliability, scalability and redundancy to prevent negative cascading regionally or nationally, a low cost goal may NOT be realistic. The key to a long term economic success is the strategic improvements of a region's infrastructure. Again, we need to embrace some short term financial discomfort as users for the long term integrity of the system and the nuclear holiday we took during the 1990's. Those were the last days that we had head room or excess capacity in many of our facilities and regions. It was also fortuitous that as over-thought and as over-spent as Y2K was, it was fortunate that the US Government was there with the appropriate amount of time and money to identify the field challenges at the data processing centers (DPC's) and zoned area management systems (ZAM's) to avoid catastrophic over heating of our waste water and cooling water with our reactors. The mini computer processing points were and are critical, single points of failure for effective maintenance. Frankly, the public sector or Local, State or Federal funds, should be available for both generating and transmission companies to meet the short and long term goals of our nation's infrastructure. It has worked in Europe and it can work here, too.

I significantly discount the 3 billion dollars set aside in 2002 and 2005 for research and investment as grossly inadequate to deal with our challenges (Clean Air Act and Clear Sky Amendment, respectively). These pieces of legislation are long on language, but are short on content. I invite the reader to form their own opinion. I was underwhelmed by the specificity of the documents. As with any important piece of landmark legislation, I looked for more details and direction rather than protracted language. This legislation anticipates the litigious world we live in and prevents or protects the government from law suits. In the private sector, when planning for and budgeting how much money and time we should spend on business continuity planning, we conduct a business impact analysis or BIA. We teach ourselves how much money we lose by the moment during an unplanned or prolonged outage and how long it takes to bring our business back up to speed. Often it takes 4 to 10 times the duration of the outage, so days of downtime may take 40 days to recover and retrieve all data. If we determine we lose 10 Million dollars a minute, then we allocate funds and time to avoid such an event. The 3.2 Billion the U.S. used to back-stop in flight trades September 11,2001 for markets which were open for less than an hour is not even a drop in the bucket for lost income, loss of life and the damage to the best known and most reliable brand in the world....The United States of America.

Let's take a look at our unique and publicly funded health care facilities. Not necessarily all of them, but the facilities where life saving or life prolonging functions are ongoing. Shouldn't there be a minimum infrastructure resiliency design criteria enhancement? The cogeneration plants in Mississippi and Louisiana were the only hospitals to survive the prolonged outages of Hurricane Katrina. Now, with eco and renewable energy conservation and creation programs commercially available and deployable, shouldn't our tax dollars pay for improvements to prolong life, help the environment and pay for improvements with 5–8 years of savings?

For those of us in the mission critical world, self help and industry guide lines are "best practices." The "third rail" of executing these plans is the cost savings challenge and maintenance programs associated with the same. Maintenance programs for government health care facilities and many publicly funded assets are almost non-existent to any exacting criteria. Nowhere in the Patriot Act, Sarbanes Oxley, National Infrastructure Protection Plan (NIPP), HIPPA or the America's Climate Security Act of 2007 (S.2191) will you find any language on criteria maintenance. In their defense, one size or one industry does not fit all. But, as we see what is spent on these reports and the money allocated to "education" going forward, we can only hope that that hospitals, government assets, transportation, critical assets, medical research, companies handling US or American citizen funds would be a bit more specific than:

- **Have a plan**
- **Document a plan**
- **Test a plan**

Take for instance our waste treatment and potable water facilities both nationally and in urban environments specifically. In the wake of regional blackouts and negative cascading of power in the west and east, as well as anomolic weather changes and events, the Department of Environmental Protection (DEP) cannot process or physically move along our human waste. This can lead to grey water within hours, if not minutes, in some urban environments. Following this, a huge bio-chemical imbalance will occur that will dwarf the cost and environmental consequences of the Exxon Valdez Oil Spill. 80,000,000 gallons of human waste is put into the East River, Harlem River and Hudson River barges and treatment facilities after a day in New York City. I saw the 14 waste stations at the water's edge as I swam around Manhattan (28.5 miles) with my son Connor in the Manhattan Island Marathon swim of 2007. We saw and swam through some very interesting things to say the least. The more interesting items were seen the closer we were to the 14 waste stations at the water's edge. But honestly, the waters are safe and wonderful. There has been much improvement over the years. However, the resiliency and redundancy of the pumps and waste management plants are viewed and treated as "non" mission critical. There are eco-friendly and mission critical solutions to address these challenges in urban environments, but the public funding is not making it to the low profile, but still critical functions and facilities.

Solar, wind, tidal or methane from waste can power these mission critical pumps, motors and systems as a decentralized solution. In this case, we would not need to not rely on central power plants or stand by 15–20 old generators with few hours of old fuel storage with little or low maintenance programs.

Eventually, this will happen. It's not a matter of "if" but "when" our urban environments waste and clean potable water networks will be mission critical. The long line of law suits to follow the poor maintenance or repair that follow a discharge of large deposits of human waste are catastrophic and simply distasteful.

The public, quasi public and private sector organizations that make these systems work daily are the unseen heroes that make these great cities effective and useful and ultimately great. These groups of the city's finest resources need now more than ever the human and capital resources tantamount to mission critical, hospitals and military installations. In business terms, we are protecting our greatest asset, our people and our second greatest asset.... our country....America

Chapter 5:

**Best Practices of Power Conservation and Creation…
In search of "Negawatts"**

The earliest and perhaps the most important consideration in outside power plant leveling, weighing, and ultimate selection is the operator. Choice or selection of the utility was less interesting before utility deregulation. The deregulation effectively divided the assets into either generating companies or wire services.

The generating companies create power and sell it. Their profits are driven by reduction in operating expenses, which are divided into human salaries and facility costs. The fossil, nuclear, and hydroelectric power that the generating companies buy is marked up and taxed, but not by excessive margins. Generating companies are weighed and scored on their uptime availability and cost, which is driven by the commoditized (and limited) source and relative demand.

If the source is in limited supply, the price will go up. Coal, for instance, is an abundant supply, but not very popular for obvious ecological reasons. Fifty permits for new power plants were recently rejected. China uses more coal for generation than any other country, and contributes more to the carbon dioxide (CO2) emissions for largely the same reason. China builds a coal generation plant every two days! In the United States, natural gas reserves are in abundant supply, but are not mined. Coal is cheaper than gas. This will change. Less expensive regions of hydroelectric or nuclear generation should be .4¢ to .6¢ per kilowatt-hour (KWH). Coal or gas will be .22¢ to .25¢ per KWH when weighted without hydro or nuclear power. The only exception is in Western New York State. These numbers are inclusive of generation and transmission. The figures are subject to change, and will likely be different by the time of publication. They often go up, but rarely do they go down. Taxes and tariffs will have a short and long term impact with pending legislation for carbon credits, proposed permits and the markets in general.

The recent and collective consensus is that coal- and gas-sourced power-generating plants are overheating the Earth. Another nagging reality is that many nuclear facilities, both nationally and internationally, will be decommissioned over the next 10 to 20 years. We are now looking for nontraditional methods of energy creation for mission-critical and non mission-critical usage that are eco-friendly and will not create more energy "well to wheel" than existing methods. In other words, some energy solutions create more waste, cost, and energy to make than to use. With a shrinking supply and an increased demand, usage is projected to go up 50% by 2031 and 100% by 2050, we can take several baby steps to find solutions swiftly, or have the courage to take larger steps on other technologies to provide self-help in power supply and power conservation in mission-critical facilities.

The commercially deployed technologies available are:

- Biomass. There are many sources of biomass, but the most commonly used are corn or soybeans to create ethanol. Blended with 15% unleaded gasoline, ethanol is commercially viable for cars, and stations are springing up. Biomass is not practical for mission-critical usage. It supplies fuel for cars and trucks, and will deplete agricultural resources and raise their prices. Biomass is appropriate and useable in collecting the methane from human and animal waste as well as decomposition waste to drive turbines for CHP technology.
- Geothermal. Heat from the earth is harnessed to drive generators and accounts for about 15 billion KWH (equal to 25 million barrels of fuel oil or 6 million tons of coal annually). It is appropriate for mission-critical uses exclusively due to its 98% of availability and reliability, to our existing systems. It is, however, more cost effective and reliable than wind or solar technologies and can be blended with other sources to secure reliability.

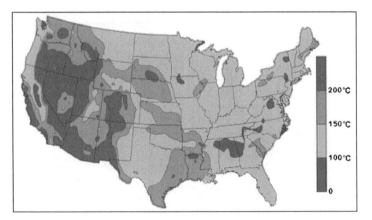

(Sub surface Temperatures)

- Wind. Wind energy is created with the movement of large blades turning generators. The challenge with wind is the high cost of installation, accidental killing of birds, and low financial return. Low-velocity wind cannot push the blades. This is not a useful solution for mission-critical needs, but useful for decentralized non critical needs. Many wind farms today are idle due to lack of maintainability or shallow depth of maintenance vendors, parts or financing.
- Hydrogen Fuel Cells: Source of fuel is abundant in water (H2O) and natural gas. The well to wheel expanses of separating hydrogen molecules from water is an energy expense and carbon footprint issue that needs to be incorporated in the matrix of go- no go.

Fuel cells consume hydrogen from reformed natural gas in most capital placed in urban environments. creation situation, but public and quasi-public capital expense programs are in place to offset day one expenses. Some natural gas bulk by agreement can be made available to reduce the cost of natural gas.

Fuel cells are stationary phosphate acid cell plants, in many cases with continuous and reliable energy creation and creating usable waste heat for energy creation.

This low and high grade heat created can be utilized (like in co-generation) for space heating, hot water operations (pool, health clubs, restaurants, laundry, etc.) and for driving absorption chillers to provide cooling. Fuel cells also act as back up facilities or base loading for facilities when the utility is interruptible short and long term.

There are over 1,000 fuel cells in place globally. This is not a near technology. The cost barrier of day one expenses was the main reason for its slow deployment. New incentives, coupled with the high cost per KWH of power, make hydrogen fuel cells more compelling for static load users.

The power plants often operate on natural gas that is reformed on site and self contained to extract the hydrogen. A thermal management system distributes the power. Any make up water required during start up and operation of the power plant is pumped into the asset.

Depending on the power plant configuration, the natural gas mixes with a small amount of recycled processed fuel and gets preheated by the reformer efforts by a heat exchanger internally. After pre heating, the fuel flows through the hydrogen sulfur where sulfur is removed. This is good!

Free of oxygen and sulfur, the fuel is pumped by an ejector and driven by pressured stream. It is all about steam again! The super heated steam is discharged as a fuel-steam mixture to a reformer where the heated catalyst provides the reaction between natural gas and stream to form hydrogen, carbon dioxide, and carbon monoxide.

Heat for this endothermic reaction is provided by burning the hydrogen in the spend fuel gas from the cell stack assembly. The reformed fuel is cooled by the heat exchanger.

- Solar. One of the more commercially deployed and a viable non mission-critical source of energy, solar power has a growth rate in the double digits. International production has improved to create and manufacture solar photovoltaic (solar PV) cells and panels to turn sunlight directly into electricity. Solar power is great for homes and non-mission critical uses. It will take the strain off existing and traditional generating companies. Solar PV sales have grown 600% since 2000 and 41% since 2006. Solar power has been successfully deployed in low KW solutions at this time.

Figure 7-1. Status of Commercial-building Energy Codes, by State

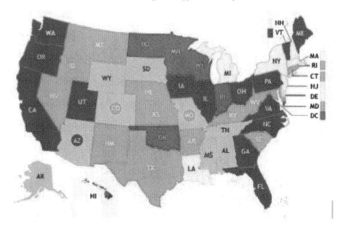

- Adopted code meets or exceeds 2006 IECC / ASHRAE 90.1-2004 or equivalent
- Meets 2003 IECC / ASHRAE 90.1-2001 or equivalent
- Meets 2001 IECC / ASHRAE 90.1-1999 or equivalent (meets EPCA)
- Precedes ASHRAE 90.1-1999 or equivalent (does not meet EPCA)
- No statewide code
- New code soon to be effective
- Significant adoptions in jurisdictions

Source: (BCAP 2007).

Figure ES-1. Comparison of Projected Electricity Use, All Scenarios, 2007 to 2011

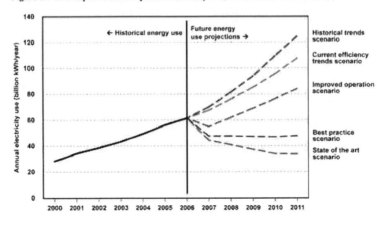

Source: EPA

- Water/Wave. Water can run continuously for 24-hour periods with fairly consistent velocity. Except for infrequent droughts, it is the most reliable alternative method of generating power and accounts for almost 10% of power generation in the United States and 75% of the alternative power generation. The water flows through dams, pushing massive turbines to create power. Other than disturbing some fish populations, this is an eco-friendly alternative with very low hazardous emissions. Water runs continuously. We have just over 850 dams greater than 5 megawatts that capture water in over 200,000 miles of rivers in the U.S. This is about 8-10% of our resource and 75% of our collective renewable source.

Gas/Coal. Gas and coal are the predominant sources of power nationally and internationally (almost 70%). We burn coal, shale, and gas to create steam or boiling water to move turbines or generators and create energy. These sources are the most damaging to the atmosphere, but they are the least expensive sources of energy. They account for about half of the source of CO_2 emissions globally. China is now the number-one CO_2 emissions producer. It surpassed the United States in 2007.

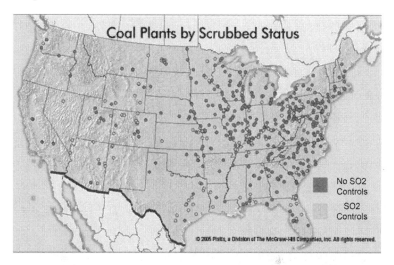

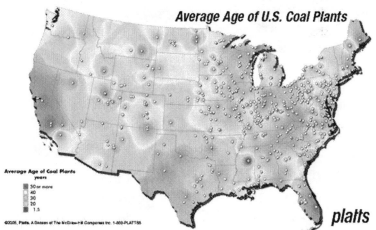

1995 SO2 Emissions

1995 SO2 Tons
80,000
40,000
15,000
5,000
1,500

2004 SO2 Emissions

2004 SO2 Tons
80,000
40,000
15,000
5,000
1,500

platts

© 2005 Platts, a Division of The McGraw-Hill Companies, Inc. • 1-800-PLATT-38

- Nuclear. Unlike energy production from fossil fuels, nuclear energy is created from the energy stored in an atom's nucleus, which is composed of protons and neutrons. The energy creation is through a process called fission. In fission, the atom's nucleus is shot by an outside neutron, which splits it apart and creates a chain reaction of uranium neutrons hitting and splitting other neutrons. Energy is generated when the heat released from the splitting of neutrons is captured and, steam is generated from water surrounding tubes. The steam turns the blades of large turbines to create electricity. The steam is cooled and stored then converted back to water for reuse.

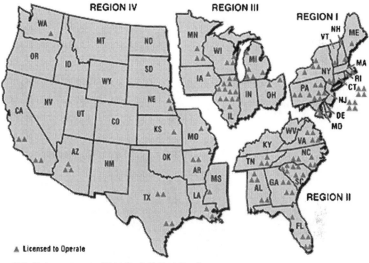

▲ Licensed to Operate

Note: There are no commercial reactors in Alaska or Hawaii.

Of the 103 nuclear plants in the United States, 24 are located in a region of drought in the Southeastern United States. All but 2 of those are located strategically on a lake or river for easy access to cooling. Do not kid yourself, coal- and gas-fired generating plants require water as well, just not as much as nuclear facilities. Some of the conduits that feed a nuclear site are 18 feet in diameter, and can run a mile long to deliver water from the deep part of a body of water to the site. The risk is that the plants will not be able to access this water for cooling and will be forced to shut down. This will not cause a power shutdown, but will likely increase the cost of power. This is simple supply-and-demand modeling. In Alabama in 2006, the Huntsville nuclear site shut down briefly. In Europe in 2006, during a drought in which thousands of people died, several nuclear plants had to shut down for about a week in Germany, France, and Spain. The volume of water required by a nuclear plant is extraordinary—in the millions of gallons per day—and it is not a candidate for surface storage. Re-piping or engineering water pipes to deeper areas away from sediment and fish is expensive and time consuming. The water access is a "gotcha" with nuclear planning and implementation. It is resolvable, however, and not a showstopper.

Close to 80% of the 441 nuclear reactors operating around the globe are more than 15 years old. The life expectancy can be 30 to 60 years with regular maintenance. However, many of the nuclear plants in the United States are coming dangerously close to decommissioning time. Nuclear power is a low-cost and appropriate means of power creation with some real, but minimal, environmental concerns. One new plant has been started in 30 years in the United States, but the world is now viewing nuclear power in a more favorable light. Outside the United States, there are 337 reactors, with 28 under construction, 62 planned, and 162 proposed worldwide.

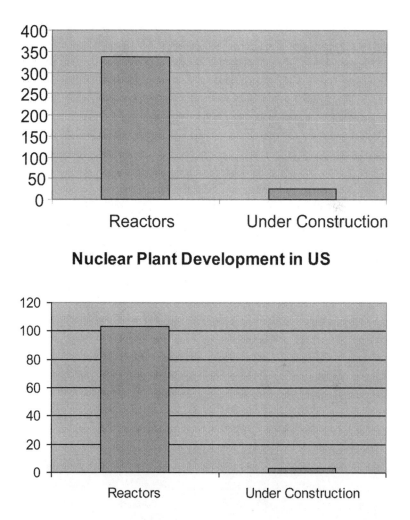

Nuclear Plant Development World Wide

Nuclear Plant Development in US

The site closest to completion in the United States is the Tennessee Valley Authority (TVA) restart of Watts Bar, set to open in 2013, and two others are expected to come on line in Texas in 2014. The United States has the most nuclear facilities in the world. The nuclear power currently contributes to about 20% of the complete global power grid. Most of these facilities were built during the 1970s and 1980s. Their 30-year licenses have been extended to 60 years, with suggested maintenance in most cases. Demand for power is outpacing supply, and energy conservation efforts are outpacing energy creation efforts. The market will price utility rates prohibitively, put higher loads on an aging and complex just-in-time network of the power grid, and will increase risk and lower reliability in most regions of the country. Only six sites have been shut down, but all will likely face obsolescence between 2020 and 2030. China is planning 15 to 30 nuclear facilities by 2020, and Russia plans 42 sites by 2030. We are not alone in the power paradigm! However, the Green Party in Germany has committed to shutting down all of that country's 17 sites by 2021 for ecological reasons. If nuclear energy is to hold its market share and keep up with high demand and CO_2-conscious states and municipalities, we need to build more nuclear plants now.

Some of the best practices for power selection and capital preservation are based on leveling components of outside plant (OSP) power rich categories are in identifying and analyzing:

- Taxes on usage.
- Reservation fees (often for mission-critical facilities as well as second feeders).
- Capital costs to build.
- Cost deliveries of primary and transmission distribution becoming de minimus.
- Negative cascading protection (engineering).

- Monitoring of network operating control centers (NOCC) are large alarms: disruptions required, human infrastructure/service.
- History of outage data, including sources in duration of outages (drunk drivers, rodents, wind, debris, ice storms).
- Equipment failure (transformer failures).
- Distance and size of substations or transmission lines to asset ($200,000 to $1 million per mile to distribute power, $5 million to $8 million to build a substation).
- The inside plant (ISP) power, right-sizing for the enterprise mainframe, storage, and mechanical components of the data center can be as complex as we want to make them. The fundamentals for what we do are remarkably simple and governed by four-function math. We can complicate the issues and cloud them with a "parade of horribles" as well as apply fears and concerns of the fluid and dynamic information technology (IT) world.
- Right-sizing the power needed, even if modular, for 15 to 20 years in the IT footprints can be challenging, but the math is simple. This is the target-rich environment for cost savings. The energy loss from the substation through the transmission lines and on to substation to the cabinet is 45% to 65%. These losses are even greater by some estimates. Like telecommunications, every time the current is manipulated by, or touches another piece of equipment, it loses efficiency. Think about it: power from a cable to a switch gear is then transformed into smaller pieces of power. It is then transformed into still smaller pieces, then rectified from alternating current (AC) to direct current (DC) to AC, then it is switched to the remote power

panel or power distribution units and then to the power strip at the cabinet.

We cannot talk about power without the corresponding input of cooling and cooling power to operate it. As power users, we are spending between $6 billion to $8 billion per year on power consumption. The cost to power services or enterprise-based devices even in low-cost footprints of $0.4 to $0.6 per KWH are greater than the cost for the device itself in two to three years. This bears repeating; the cost to power/cool the device is more than the price of the device itself in the short term. In IT terms, that means there is more money in software than hardware. Something seems very wrong about that to me.

If we do not figure out how to utilize the waste product of power (heat) or reduce the creation of the waste product, we will be in big trouble collectively.

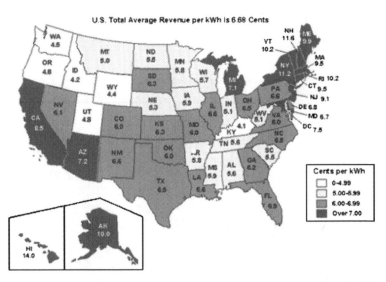

U.S. Total Average Revenue per kWh is 6.68 Cents

2005

Cogeneration or CHP in a decentralized topology will distress the grid, and slow but not stop the centralized power plant "refreshes" or new build requirements. The price of natural gas has an obvious impact on one source of fossil fuel to the central or decentralized solutions. Methane gases for micro grids are greenhouse friendly decentralized solutions for residential and commercial needs inclusive of cooling requirements.

Cogeneration or natural gas solutions work best as centralized and distributed solutions when the tarriffed footprint costs per KWH are 14-16 cents or greater. The rising cost of natural gas will creep into the cost per KWH, by 6-18 month increments due to long term gas contracts with the utility as well as their source diversification of nuclear, coal, hydro and other sources.

Another discouraging component for the coal gasification and natural gas solutions of cogeneration as a primary or supplementary energy is carbon capture and sequestration or (CCS). The prevailing regulations are speculative and are organized "guess work." The most recent effort of CCS is a legislative memorandum being circulated in Congress called "Moratorium on Uncontrolled Power Plants Act of 2008." This proposes in the second quarter of the moratorium to sequester 85% of CO_2 emissions. This new Act, by the way, will reduce the energy effectiveness from 7–12%! That is about 80-100 megawatts for larger plants. That would be a huge loss financially to the generation companies, forcing them to build more, or rely on technology load shedding techniques in transmission/distribution to just keep a zero sum game and not go backwards. Plant decommissioning due to age and ineffiencies and human/electronic needs growth will continue to challenge the supply and demand standards.

Chip makers and vendors are paying special interest to the problem. The "Green Grid" is made up of big-brand power users: IBM, AMD, HP, Microsoft, AMC, and others. Their goal is to create efficiencies and standards. If the situation is analogous to what happened in the telecom world, they have had to move from Centrex or DMS, 5 E switches to IP or "soft switches" and cannibalize successful business lines to "get green." Then, we will have to wait three to five years for commercially deployed solutions other than those discussed in these pages.

Chapter 6:

Data Center and Mission Critical Power Needs-
The Impact of Moore's Law

The Department of Energy recently released the national supply and demand for energy creation and consumption in America. The DOE indicated that mission critical power demand is currently 1.5% of total demand. They later revised that number to 2% in 2008, growing at a 20% per annum increase year after year.

These numbers are consistent with mission critical equipment supplies I work with, as well as the information technology IT kit provides for year after year growth. The insatiable technology appetite for processing voice, video and data is extraordinary. The newly mandated storage requirements of Sarbanes-Oxley, HIPPA and the Patriot Act have figured largely IT in the compliance game of mission critical document management.

For the data center of the 1980s and 1990s, growth was predicted at 50% to 70% over 10 to 13 years. That means, if users needed (day 1 and future) 100,000 square feet of white space, they would need to plan for 150,000 to 170,000 square feet total inclusive of environmental to support the white space and levels of redundancy. (Today's load demand 2.0 to 2.5 times the white space for the total space required.) That figure was based on existing velocity of growth for white space. Power densities were 15 to 25 watts a foot, and cooling was fairly static. Moore's Law, although in place at this time, did not anticipate the more powerful chips associated with cooling configurations until years later. More recently, particularly following the drama of Y2K and the terrorist events of September 11, there has been accelerated growth and interest in large data centers. The "mega–data center" has lost its place 90% of all data centers are under 10,000 square feet. The cost and time required to design the 20,000 to 100,000 square feet of white space was and is overwhelming. It often made sense to augment human space or office space within a strategic asset, rather than taking the time and expense to secure, design, improve, and maintain a larger data center. Operating expenses over 15 to 20 years are staggering, and are the justification many users give for not designing a larger data center. Data centers are not their core competency or core business, so why should they pour $3 to $10 million a year into just the maintenance of a noncore business? Below is a typical layout of a two story multi tenant data center. Users share infrastructure a la carte to avoid higher capital expenses and have resigned themselves to a higher operating expense.

More recently, and largely in the reaction to the World Trade Center attacks, various white papers and regulations regarding data centers have been taking a more centralized approach from a human, equipment, and a real estate cost point of view. Information technology (IT) service topology delivery inspired a move to a more centralized model.

Collapsing multiple sites into fewer sites is a target-rich environment for cost savings. Some of the reasons to migrate out of existing or legacy "spoke-and-wheel" data centers are:

← They are too close to the primary facility.

← The infrastructure is outdated in power, cooling, and telecom infrastructure and is no longer able even with retrofits to satisfy new power and air-conditioning requirements.

According to Gartner, "server rationalization, hardware growth, and cost containment" are "driving the consolidation of enterprise data processing sites into larger data centers." Underutilized and oversupplied servers became financially imprudent to manage and maintain. Gartner discussed the rise of distributed computing and other trends, which led them to the deduction that large data centers were on the decline. The rise of distributed computing and other trends drove into decline large data processing sites that characterize the era of mainframe dominance. Now, however, data centers are rising in importance. There is a real relationship between lost revenues and downtime. The consumer can see a relationship between the negative cascading of power in California, the regional outage in New York, and the events of September 11. When these events are combined with corporate governance concerns and newer legislation, companies recognize how much money they will lose by the moment in the event of a catastrophic or prolonged outage. There are a number of main drivers associated with interest in larger data centers, they include server rationalization, cost containment, improved security and business continuity (a new corporate discipline), growth in hardware, and containing software. Current reasons for unique focus and interest in larger data centers include:

← Now more than ever, users are looking for IT solutions to reduce human costs or overall operating expenses, effectively trying to create their own "special sauce" to become more efficient in data processing, trading, clearing, and storing data (cost containment).

← Revised and ongoing, there has been continued interest in reducing the footprint of various legacy or antiquated data centers. As a result, many users have reduced the number of sites and placed equipment in larger sites.

← There is a need to improve security and create a business continuity plan (BCP). No longer is triple fail-safe security satisfactory for most data centers (triple fail-safe, human, closed-circuit television, and proximity). Corporate governance is establishing a new level of criteria for possible IT, intervention, human intervention, and cyberterrorism. It is common sense to recognize that with fewer assets to protect, there will be a smaller risk of interruption.

← Another reason for the growth in larger data centers is the hardware requirements. In the past few years, we have seen large server deployments. In 2006 and 2007, blade or multiple server deployments increased by 55%. The multiple servers were rolled into data centers and were often underutilized. Combined with server deployments, the storage ability of new solutions for various equipment providers has increased significantly, which requires physical space. For nonfinancial and financial companies, storage has been tantamount to productivity due to the Health Insurance Portability and the Accountability Act. These are now generally accepted accounting principle

(GAAP) requirements created by the Sarbanes-Oxley Act. Records need to be kept for a minimum of seven years. Implementation is slow. In 2007, I went to an emergency room and was provided with two pages of 30 detailed name and address stickers for various files. My visit was recorded with over 30 user groups or companies, providing a redundant paper trail that is both cost inefficient and a waste of real estate space.

The reduced number of data centers and a company's willingness to invest in multiple footprints have resulted in fewer and more meaningful data centers. These data centers are now often broken into simple squares, or rectangles commonly referred to as "cells," "pods," or "data halls" to satisfy redundancy requirements inside the plant or inside the envelope. The topology of the mega–data center or multiple data centers has been reduced to an active-active (both in the same region), active-passive (one outside of the region), or geoplexed footprint scenario (out of region and out of the likely area subject to an Act of God).

These new, enlarged data centers aim to scale to size the white space and power/cooling needs as well as the human infrastructure to enhance functionality of the hub or server locations within the organization. The bigger data centers also are centers of extraordinary human infrastructure and best practices to be deployed elsewhere around the country. One of the goals is to have the least amount of redundancy of hardware and infrastructure across multiple sites by concentrating capital, human intellectual capital, and operating expenses into a smaller number of sites, and then making those sites as meaningful and redundant as reasonable. Not only can hardware costs can run rampant, but unused software licensing costs and taxes can run into the tens of millions of dollars. By leveraging buying power and considering the economic incentives of new deals for sales taxes for kit, utility, and telecom transmission costs can be contained. This kind of model can be adapted, with the result scaled to create the most cost-effective, cost efficient deployment of physical and cyber assets. This is important for corporate executives and real estate brokers to leverage the consumer's spend. There is no reason for users to occupy a footprint where the cost per kilowatt-hour (KWH) is $0.13 to $0.15, or even $0.22 per KWH and there is full sales tax on equipment when users can move to a footprint of $0.3 to $0.5 per KWH with limited or no sales tax. The difference is staggering for large users, it can mean $300 to $400 million over 20 years.

One technology variable entering the market is the container data center. IT vendors are creating single source containers of IT kit, power distribution and cooling to satisfy day one and future modular solutions in enterprise, storage and mainframe needs. The technology does not have the field time validation as of yet, but good pre market press. I call this a great "white board" and board room sell. The TCO models are all vendor favorable and more information is required. Field implementation of same can be difficult. One thing is clear: to get between the wall and the wall paper of profitability of facilities and IT utilization, some fundamental paradigms need to change. IT needs to be more closely involved with monolithic data center designing and building. Energy solutions and eco-friendly power distribution are required and the waste product of energy...heat needs to be reduced. This will also reduce a mechanical needs and loads. Power to the application needs to be modeled and deployed as a TCO template for effective data center deployment and the "fur ball" of critical and non-critical applications can and will be physically and virtually separated. It is also best to place the physical IT assets in parts of the world with low or no sales tax, low power (3-5 cents) and telecom transmission expenses. Anything less than the above in "all" or "partial" measures and the human infrastructures supporting mission critical data are under funded or acting like lemmings out of lack of knowledge.

Another compelling reason to go to the large or main data center scenario is the use of virtualization to improve asset utilization and virtual capacity planning. Quite often user efficiency is somewhere between 8% and 25% of modeled capacity. Emerging virtualization technologies, particularly in server and storage equipment, offer the best asset utilization potential. "Solutions providers" are growing at an extraordinary rate, currently there exist between 240,000 and 260,000 providers. These data center operators "manage your mess for less," according to ad campaigns. Their average time in the business is just over 10 years, the average duration of client relationship is almost 8 years. The average number of customers is 180. Solution providers manage a balanced menu of services:

- ← Hardware: 26%
- ← Software: 25%
- ← Services: 49% (everything else)

Solution providers velocity of growth is good. For every three clients they gain, they lose just one. This is a positive churn and indicates that there is some product loyalty, which in a commoditized world is good for all of us.

These hardware and software technologies can also improve operational and operating expense processes, driving down telecom, human infrastructure, and hardware costs. Although virtualization does leave some risk in terms of inflating data, it also provides a meaningful cost savings and footprint savings scenario.

This server proliferation or virtualization has helped the IT industry to shift from higher-priced mainframes to lower-cost servers. It also has contributed to an exponential increase in the number of multiple servers deployed by financial institutions and ordinary user groups.

The blade surge has had an impact on Value-Added Resources (VARs) in the deployment and utilization of the kit:

← 14% of VARs sold blades in 2007.
← 20% of VARs plan to sell blades in 2008.
← Represents a 45% increase for 2007 leading all technologies.

The blade server shipments have been recently documented to the measure growth and velocity, which has caused a buzz regarding the environmentals. (Note that several customers and clients are moving toward the multi-"U" topology and away from the blade and heating and cooling challenges.) Projected growth of the blade technology and pricing is:

← 2006: 620,000 blades shipped—average selling price $4,189

← 2007 (estimated): 856,000 will ship—average selling price $3,967

← 2011 (estimated): 2.4 million will ship—average selling price $3,605

In comparison, overall server shipments were:

← 2006: 7.8 million

← 2011 (estimated): 11.3 million

This has also been combined with customer's historical interest in deploying a single special application over a server while not risking other encryption over another critical application. This is like a family of four being the sole occupants of a 40-key hotel and living there indefinitely. Virtualization techniques and applications have compelling financial and human infrastructure reasons for adoption, such adoption is taking place only slowly, however.

The server processes have continued to evolve according to Moore's Law, and they continue to double in density every 18 months. Although the benefits to the IT user groups have enabled them to run bigger-bandwidth applications, and to work in batches, they have also resulted in a massive increase in power consumption due to the more powerful chips, cooling requirements and environmental expenses. Keep in mind that as we discuss the newer, larger data centers, and the unique time and capital effort it takes to design, build, and maintain such unique facilities, that they are built with Tiers 3 and 4 in mind. A brief discussion of the tiering models is worthwhile so we know what the objective is.

Tier 1. Single path of power and cooling. No redundant components. Less than 28.8 hours of downtime per year (satisfactory for noncritical users, form infrastructure requirements for most telecoms).

Tier 2. Single path for power and cooling distribution and redundant components. Less than 22.0 hours of downtime per year (common design-build scenario for telecoms post deregulation, insurance companies, credit card companies, and media outlets).

Tier 3. Multiple power and cooling distribution paths, but only one active redundant component currently maintainable. Less than 1.6 hours of downtime per year (traditionally military specifications grew out of the enterprise and the mission-critical phenomena post deregulation).

Tier 4. Multiple active power and cooling distribution paths, redundant components, all tolerant. Less than 0.4 hours of downtime per year. In Tier 3 and 4 scenarios, our architectural suggestions are not offered. These assets can actually be vertical. Except from the conventional wisdom of making these tiers horizontal, there exist compelling reasons—namely cost savings and efficiencies—for having the white spaces on the second floor and the environmentals to serve the white space fed from directly below the white space.

This provides an easy-to-understand summary of the tiering differences commonly discussed today. Most of the data center world is working between a Tier 2 and Tier 3 infrastructure for concurrent maintainability goals. Tiers 1 and 4 are becoming more and more uncommon.

The most important point about the Tiering models is the difference between Tier 2 and 3. This is a defining moment. The price difference between these tiers is significant and largely due to the fact that Tier 3, by definition, is concurrently maintainable. Concurrent maintenance systems require the system to be shut off in portions of the asset so a certain area can have an anticipated outage, an unanticipated outage, or a predicted scheduled maintenance. Therefore, the incremental investment for dual electrical mechanical systems to meet the concurrent maintainability and fault-tolerant criteria causes a significant increase in capital expense. Tier 1 and Tier 2 are practically linear. There is backup for anticipated outage, but bypasses and some redundancies in general give them their linear design.

The costs and the cost benefits of the various Tiers are fluid. In addition to modified descriptions of Tiers like "Tier 4 light" or "Tier 3 on steroids," the pricing gymnastics are often "real time" with scheduling and challenging lead times and interests. In effect, the cost of copper has gone up 100% in the last five years. Believe it or not, materials go up approximately 1% per month. They never go down, but they sometimes remain flat. Therefore, costs of switch gear, uninterruptible power supply modules, cabling, and labor have increased significantly (6-13% per annum). The cost of lead has increased similarly, so wet cell batteries are now remarkably expensive. See Exhibit 6.1 for terms of pricing and tiering.

Again, if the shelf life of this book is three to five years, we anticipate that the pricing models in the exhibit will be outdated within just 18 months. In 2006, China bought 80% of the world's supply of concrete. Do you think that had an impact on supply and demand? Currently India and China consume 80% of all energy, and only half of the population of these countries has plumbing and lighting. China builds a city containing 8 million people—a city equivalent to the size of New York City—every three years. China's population is literally dropping their hand tools and walking to the cities in order to work. The point here is that renewable energy and finite resources are impacting our ability to site, design, and build mission-critical facilities.

Aside from the capital expenses involved in building a larger data center, the operating expenses are equally onerous. As most data center managers will explain, beyond software costs, their major concern has been driving down operating expenses. Data center managers are trying to achieve better utilization by driving down utility waste from electrical distribution and cooling. To satisfy the explosive growth of the enterprise and storage environments, an extraordinary number of raw processors and supersized chips on the data center floor and disc space capacity are being managed within the footprint. This unique growth has come at an operating cost that is a major focus of the industry. Beyond the hot aisle, cold aisle, and hot spot seminars held around the country every quarter, a unique focus is now bringing down the operating expense of the data center.

Total cost of ownership (TCO) models once reserved for the hardware or IT discipline of the data center environment have migrated into the facilities. Once again, the private sector is looking to the public sector for guidance and forecasts regarding the extraordinary power usage of these data centers. It is fairly clear at this point for consultants like myself to sort out various parts of the country with acceptable tolerances of Acts of God and human intervention. We also need to identify level and score regions of the country with meaningful, reliable, and relatively cost effective power distribution (not only the capital expense to build, but the operating expense to maintain same). Following most deregulation scenarios or breakups, the first operating expense to be discounted is maintenance and management. The first thing to leave the utility sector was the human infrastructure which once maintained this critical footprint (generators, transmission, and substations) now between 30 and 60 years old in many parts of the country.

Unfortunately, it comes down to a base level of maintenance of rights-of-way, tree trimming, snow removal, upgrade software controls, and network operating control center maintenance storage of spare parts and management. Ask a utility the costs of primary and reserve power, the number of transformers at the substation, and the source of transmission power. Location of the spare parts has become more important than history of outages. Who manufactures your spare parts? Where do you warehouse them? How many trucks do you have? Can I go up in a helicopter and identify your rights-of-way? Allow me to level and score your tree trimming and maintenance procedures.

On the heels of deregulation and due to the explosive growth of the data center market segment, President Bush signed into law HR 5646 on December 8, 2007. This is in order to study and promote the use of energy-efficient computer servers in the United States. The Environmental Protection Agency (EPA) worked with computer makers to recommend that the government adopt new incentives to handle the problem of rising power consumption and computer data centers. The EPA had six months to submit the study. It determined that power consumption for data centers unique to mission-critical environments accounts for approximately 1.5% of our total usage in the United States. This is equal to the annual consumption of the state of Michigan, or the energy for every television in the country to be run at the same time.

These mega–data centers generally are sited in rural or "cornfield" scenarios on the bubble of urban environments. These locations are chosen because generally that is where the confluence of meaningful power, smart humans, manufacturing, legacy factory, and/or distribution centers meet significant telecommunications (fiber optic scalable, burstable, and synchronous optical networks [SONET]) networks to service multiple humans or commercial development. Managers of these data centers want to site them near urban locations. These cities can not be too rural (cornfield Scenarious). The utility needs of a 30- to 60-megawatt data center will be similar to the power used to light up a small town. Although data centers have few well-paying jobs, they provide hundreds of jobs (up to 300 at times) with evergreen construction and vendor support on-site or nearby. The greatest benefit to the local and state government is in the sales, real estate, and personal property taxes, which vary dramatically from state to state.

Data center equipment is composed of hardware or servers combined with the software to make them cost efficient and profitable. Analysts expect the server market in the United States to grow from 2.8 million units, or $21 billion, in 2005, to 4.9 million units, or $25 billion, in 2009. This is an increase of almost 50% in five years, according to a recent IDC forecast. This is consistent with Moore's Law. The history and future of Moore's Law and Gordon Moore's legacy are well known and worth reviewing.

In 1965, Intel cofounder Gordon Moore predicted the economics of the next four decades of computer power. His theory was tucked away in an essay in an April issue of "Electronics" magazine. (An original copy of this issue recently sold for $10,000.) Gordon Moore said that the transistor density or integrated circuits at minimum would double roughly every 18 months. Over the years, there have been spikes and dips in the model based on the economy and commercially deployed equipment. However, today, what has become known as Moore's Law is commonly taken to mean that the cost of computing power doubles every 18 months. The cornerstone of what Moore articulated in the article has drifted into all components of the mainframe, enterprise environments and trends in hard disc capacity over the past few decades—our ability to manipulate and store data. Commercially deployed technologies of broadband, video, audio, and fat bandwidth combine to satisfy the thirst for processing force growth in processing power. The commercial analogy I use for the increase in chip breakthroughs is the trickle-down benefits of adjacent processing in PDAs (personal digital assistants) and cell phones that were made with battery technology. Eight to ten years ago, our phones were small and included multiple functions, but when fully utilized they only lasted 15 to 20 minutes. When batteries became lighter and lasted longer, the PDAs and phones began sell more. This is the same with the enterprise environment. When the chips became faster, technology grew cheaper and therefore became more commercially viable. The good news here is that the cost to manipulate data or encryption is now coming down in price. However, the cost to build Tier 4 environmentals has gone up nearly 100% in the last ten years. One noted industry consultant has indicated that the cost per processor has fallen 29% per year in three years. As a result, the IT budget will buy 2.7 times more processors and 12 times more processing power. However, the cost to manage the environmentals—cooling, fire suppression, and power distribution—has gone up 20% to 35% over the same time frame. We are collectively being forced out of our comfort zones of the AC uninterruptible power supply power plants and cogeneration to slow the "hockey stick" price increase models for Tier 3 and Tier 4 environmentals.

Recently, a study conducted by the Data Center Users Group found that 96% of current facilities are projected to be at capacity by the year 2011. This opinion may be guided by the fluid dynamics of hardware, software, and virtualization. These predictions are not a straight line. Technology simply moves too swiftly.

The life cycle of the data center is shrinking. Aggressive corporate growth plans, along with the use of new, more powerful server technologies, are pushing data centers to their limits. A host of facilities in urban environments and legacy facilities designed for 35 to 50 watts per square foot are maxed out. Although these centers may have the physical space to add more equipment, they lack the power and cooling capacity to support that equipment.

The challenge of consulting the super-data center dynamic of how much, how long, where, how to make the data center scalable, flexible, and versatile is extraordinary. However, to "future-proof" what we call "rocket ship real estate" is challenging and often protracted. Consultants need to take a look at the velocity of growth of hardware, add-ons, and utilization at the cabinet. The cornerstone of what Moore articulated in the article has drifted into all components of the mainframe, enterprise environments and trends in hard disc capacity over the past few decades—our ability to manipulate and store data.

A relevant history and velocity of growth is critical before a user gets over-vendored into overdesigning an infrastructure based on organic and anomolic growth patterns.

Exhibit 6.1 Models for Growth

	Tier I	Tier II	Tier III	Tier IV
Augment existing 50-100,000 sq ft of white space	50 watts $650 psf	100 watts $850 psf	100 watts $2,700 psf	150 watts $3,500 psf
Colocation (up to 20,000 sq ft)	$150 psf per annum w/o electricity & setup	$350 psf per annum w/o electricity & setup	$550 psf per annum w/o electricity & setup	None available
Greenfield 50-100,000 sq ft of white space	$650 psf	$1,200 psf	$2,200 psf	$2,800 psf

At this point, the steering committee can make contributions as to where they think the company is going. The only way companies are going to overcome the obstacles of running out of power, cooling, and footprint is by designing for higher densities and employing adaptive IT infrastructures that will have greater flexibility to adapt to industry changes.

I am not necessarily an advocate for the 200- to 250-watt-per-square-foot environments. Except for exclusive, super-high, or compartmentalized environments, these are not commercially viable. In these small, super-high rooms, if one cooling component fails, the environments will overheat within a matter of seconds, not minutes. Generally designers of such data centers install additional air-handling units for the "just-in-case" scenario. Thus, they effectively use up the floor space that you expected to "save." To move to super-high, super-perforated tiles that you anticipated to use in front of your cabinets effectively have a speed of about 125 to 175 miles an hour, which eliminates all (dresses) in the environment, and goggles, not glasses may be standard issue. These data centers are not necessarily practical. Spreading high loads over low-load densities appears to lessen the risk of this model. Furthermore, these densities are rarely met. One large utility in New Jersey has not seen densities over 65 watts per square foot, one large data center landlord has seen densities between 45 to 65 watt per square foot. Let us assume growth is forthcoming. Let us provide room for equipment growth, but let us be reasonable!

The super-high scenarios are being commercially deployed with not only a hot-aisle/cold-aisle configuration but with a spreading of load. Until recently, data center managers wanted to line up servers, mainframe, and storage networks by discipline or by faceplate. This may look good, but it is not efficient in reality. Most data center managers do not know, within 24 hours of a move, what equipment is coming into the center and what the new loads will be. Circuiting and conductoring are often done "on the fly by" in-house electricians. This is not a best case scenario, and can lead to trouble.

Rolling out equipment like soldiers is not commercially viable. A 0.5-kilowatt (KW) cabinet should probably sit next to a 10-kW cabinet to blend the heat distribution. Beyond 6 KW (by average per cabinet), unique cooling solutions need to be provided.

Because most data center users do not know what equipment is being rolled into the environment, this means it has generally not made it through the lab or the testing bins to apply the special applications. The speed to market has eclipsed data center functionality-how close to substation, fiber, or outside plant issues. Effectively, whips and power circuits with cooling are expected to be in place to anticipate unknown server distribution. This ongoing turf war duplicates the friction between the IT and facilities groups which has been going on since the 1990's. More functional and effective companies have put down their weapons, eliminated the silos, and learned to work together.

One financial institution that I have worked with was the biggest blade buyer in the country in 2006. On average, it was receiving 30 blades per week. Although the utilization of these blades was under 15%, they were being commercially deployed to satisfy user groups that did not want to share resources. Once fixed in place, the blades would grow vertically (within the chassis/cabinet) and not horizontally, which would take up more floor space.

This is a more mature view of growth from the point of view of load density and balancing. Equipment utilization is the cornerstone of effective data center management. Getting business units to share physical equipment and take that back to the profit and loss statement is far from easy, but it is well worth the effort and associated savings. Companies are now working hard to reduce operating expenses without reducing or compromising the integrity of the plant.

Chapter 7:

"Farmatopia" or "Farmageddon"

I have taken liberties with creating new words for our current agricultural energy opportunities. "Farmatopia" or "Farmageddon," can be considered our green and renewable energy sources. These sources include solar, wind, geothermal, tidal, wave, and micro dams, etc. They are moving forward and emerging, but too slowly for the commercially deployed velocity of construction to have a noteworthy impact on our economy. The eco-friendly energy creation methods and associated parts per million volume (PPMV) carbon reduction in the short term have long sale cycles in the commercial markets and longer cycles in the legislative or incentive sectors of the markets. There will be little or no noticeable impact on the existing US power grid with green energy sources to support our increased demand, doubling of the population by 2040 and the concurrent decommissioning of various generating facilities around the country. There is no stop gap or "step in" the green energy "silver bullet" program to support our US or regional needs as our national resources expire. This is the one thing both Democrats and Republican can agree on- except for those who are truly off the reservation.

The long term consideration is not about OPEC's ability or willingness to supply oil to the US, China or India. It's not a discussion of what happens when gasoline reaches $10 or $15 per gallon, and how much of that goes to the oil companies. We have a finite or limited supply, not only of liquid petroleum resources but all fossil fuels.... period. Ethanol and alternative fuels continue to lobby for massive subsidies or tax breaks, and are getting them at the expense of other produce and land uses.

Some opponents argue to open up the Antarctic region where 25% of our resources are hiding under our melting masses. This has been pretty much off limits for conservation reasons. At best, it would postpone the reality of maintaining our finite supplies or putting off our dependence on other resources. Some folks consider this movement or theme has become "code" for "drill it all." The reality is that is unbelievably wasteful that we actually burn oil with all of its usefulness. Our 400 million year old treasure literally goes up in smoke! It took nature hundreds of millions of years to create fossil fuel by the decomposition of organic materials, and we have only so much of these. Nearly every energy innovation we have come up with over time has been a low cost solution from transportation to generation and distribution. Today, and in the foreseeable future, our energy solutions will NOT be low cost solutions. Compounding this reality is an increasing doubling population, more challenging field conditions and fewer natural resources. This is not a soothing conversation about the endangered polar bear or migrating tree beetles in Alaska killing our CO_2 absorbing greenery. This is a very serious and real look at our national security, energy security and not energy independence. By the way, five polar bears die a year due to ice melting and 50 die due to poaching. Let's stop the poaching if the polar bear is the issue. To protect our food, resources and economy, we need to look to ourselves to solve the problem. With only 10% of the world's petroleum resources NOT controlled by state run companies, we are at risk of keeping the social and economical barometer or the human infrastructure in place. For transportation and other uses, unfortunately, ethanol is being pushed as a major solution to our current challenge. Blended with gasoline, it makes a compelling "board room" solution. It is green, it is here, and we can grow and replace it. Rather than spending more time and money on bio waste or agricultural waste, we dove head first into the heavily subsidized lobby of the corn based ethanol producers. This technology has proven inefficient in the "well to wheel" cost and CO2 emission reductions goals. This has made the "super farmer" very rich and the smaller farmer (under 300 acres...not so much). A global increase of 30% of production and fuel has resulted from the shift to energy farming. Recent shortages and Acts of God have contributed to the increase in global expense for goods, but the impact has been mostly related to sugar based crops. One study shows a 75% increase in foods. This is well over the 3% reported by the US. Alternative fuel farmers to sort out the incentive targets and protocols for farming of today and the future. Switch grass, soy, sugar cane and now "jatropha" have extraordinary bio diesel/fuel applications currently and internationally. It is switch grass that gets most of the headlines, but jatropha grows faster.... it is planted once over 36 years and has an oil base that rivals palm oil. The plant/tree grows in the most unfriendly climates (Africa and India) with little maintenance. Large plantation owners will make a fortune growing hectares (1 hectare = 2,500 square acres) of jatropha. One acre is about the size of one of the building floors. Currently 5,500 farmers have planted 11,000 hectares. 10,000 are planned by the beginning of 2009, but the carbon credits for the "electrification of rural land" as a CDM under the Kyoto protocol will provide vital credits to make the business plan viable. This is a good alterative fuel. Jatropha is not fit for human consumption and therefore will not impact food prices globally. Cooperatives of interest are required to harvest the brain trust and capture the credits. Don't forget, we have to get this fuel to market and in the stations to tanks. All those "well to wheel" costs do not go away. As well, we are still burning a fuel with fewer particulants, and not zero particulants. We collectively are creating more distance from OPEC, but are not divorcing ourselves from OPEC. However, it's a start, and to stay in the solution and not the problem side of the equation. We currently have ethanol tariffs in place from South America. We also have sugar cane sources to protect our production, and now we have a corn and corn syrup shortage that is impacting the cost of many goods and services here. Unfortunately, the corn syrup/product industry rules here in the US rice-based glucose are out of the market without the government subsidies that make high fructose products commercially viable. Call me crazy, but if we could ban or reduce high fructose corn syrup as a food additive, we would be healthier and the price of bio fuel would go down. Our fast food would taste like hell until we got used to rice based products or other alternatives, but corn-properly used, subsidized and consumed would take its place in the "Darwinian" collective food/ energy chain quagmire we find ourselves in today.

How did we get here? From a liquid fossil fuel or gasoline perspective, the Department of Energy indicates this major source of transportation fuel consumed over 140 billion gallons last year. This is in stark contrast to 50 billion gallons of alternative fuels (methane, ethanol, petroleum gas, natural gas) plus 6 billion gallons of additives like butyl, tertiary, methyl, ethanol and oxygenates.....these are NOT renewable.

Total fuel consumption has risen 28% over the past 10 years. Alternative fuels have risen 62% over the same period. The fuels are making progress, but are still light years away in terms of supplements that are needed for a transition from fossil fuels. At this rate, the world will just plain run out or organic fossil fuels.

The sources of fuel or oil reserves are as follows:

Saudi Arabia-20%
Canada-14%
Iran-10%
Iraq-9%
Kuwait-8%
U.A.E.-7%
Venezuala-6%
Russia-5%
Nigeria-3%
Libya-3%
Kazakhstan-2%
USA-2%
China-1%
Qatar-1%
Others-10%

Where we get it. The following is where the US imported oil from 2007:

Canada-19%
Saudi Arabia-15%
Venezuela-12%
Nigeria-11%
Iraq-5%
Angola-5%
Brazil-2%
Colombia-1%
Others-10%

The above does tell the current story of who has it and where we get it, but realize that transportation CO_2 emissions and transportation energy needs are only 14% of the world's collective "need." Energy creation or power amounts to 25% of the world's collective "need."

We should want to encourage energy conservation and energy creation improvements wherever the "well to wheel" effort and expense does not have a gross negative impact on the earth, but the savings category for natural resources or fossil fuels is in energy creation.

With 10 electrical consortiums of:

- WSCC-13 states West Coast and Canada
- MAPP-8 states West, North and Midwest as well as Canada
- SAA-4 states South Midwest near Texas
- ERCOT-most of Texas (of course)
- ECAR-6 states Northeast-Great Lakes region

- SERC-9 states Southeast
- FRCC-most of Florida
- MAAC-3 states, NJ PA Del
- NPCC-5 states Northeast and Canada

The lowest reserves of energy by consortium post regulation are in the Northeast. Pre deregulation of 20% or greater reservation of energy reserves was required. Post deregulation, the following is the current truth:

2002	2008
ECAR—20%	14%
SERC—20%	13%
MAAC—20%	14%
NPCC—20%	16%

Three of these grids provide most of the U.S. capacity which is made up of over 1,000 power companies growing at about 25% every 10 years for gross production.

The green or renewable energy accounts for only 1 to 2% (depending on whose data points you are looking at) of the total energy generation for decentralized usage or to supplement the existing networks.

So, what are the real, proven, commercially deployed "green energy" or low carbon emissions energy creation sources worthy of deployment, incentives and usage?

- Biomass
- Hydrogen fuel cell
- Geo thermal
- Hydraulic
- Solar
- Wind
- Wave/Tidal

Biomass- is most often trash, and sometimes stock feed. When burned or gasified, it creates energy. Trash dumps are great resources for methane. Ridged pipes are inserted in trash dumps and gas escapes, or domes can be built over waste and gas is captured and manipulated. Biomass is the result of the excess parts of stock feed, or undesirable parts of grain crops harvested for food and burned as fuel in power plants.

Geothermal- and hydraulic energy are renewable. Other renewable energy sources are wave, tidal, micro dam, and thermal vents (land or ocean). The fundamental principles for energy source is that earth's free flowing energy is captured and controlled to move blades of turbines or similar device in synch with generators to create power. Geo thermal and hydro powered facilities are high visibility. They draw the attention of Local, State and EPA entities to guide, enhance, and hopefully expedite the protracted siting, designing, construction and commissioning process while disturbing the earth as little as possible. The Hoover Dam powers Las Vegas, California, and Arizona with its 17 main turbines. The Colorado River in Texas has six major dams on the Colorado River which provide power to millions of people via 10 electrical co-ops and 35 cities. This is equal to half the water volume of 100,000,000 gallons per minute of the Niagara falls that is diverted toward energy creation to be split between the U.S. and Canada. Tidal power has only been harnessed in the UK for only about 1,000 years. Wave technology harnesses the vertical movement of the oceans waves, wind, and tides. This is also subject to the spikes and lulls the ocean has naturally.

The other source of power generating much media attention and West Coast decentralized and centralized commercial viability is the solar PV panels (the "Solar Grail"). Solar energy is available in about 70% of the United States. Having been born in "Syberacuse" (Upstate New York), a city with the third fewest sunny days in the US, I am somewhat skeptical of its large scale deployment, commercial viability and physical deployment. Even worse, I went to college in Buffalo, New York, a city with nearly as many "snow days" as "Syberacuse." Great city if you like lots of snow. However, the reality is that several hundreds of acres of land are being improved for the centralized creation of solar power. In one plan, the company is using rooftops of warehouse buildings which from a maintainability and installation point of view, I question. But be that as it may...I applaud the effort. Storage and connectivity to the Grid is still a challenge. The current conversions of solar energy to electrical electricity are a bit inefficient on the large scale, but not on the domestic or small scale. Conservation deployment, it's a huge success for the consumer and sheds load off the incumbent grid.

Wind turbines are a green energy source with the same issues as wave technology. What is odd is that over 90% of green energy construction spending this century went towards wind farms. This tells us a few things. One, they are not cheap. Two, they have a tremendous psychological appeal. Wind turbines are expensive, large, and have the same challenges of dips and spikes that the irregular earth's atmosphere creates, not to mention energy storage. US policy dictates that 5% of our total energy output to come from wind by 2020. This is just around the corner, since it takes 3-6 years to commercially deploy these systems and a lot of capacity to consider!

The largest wind plant in the world is in Wallawala, Washington. This plant has 396 turbine generators to serve over 50,000 homes. North Carolina is building the largest plant currently. At 300 ft tall, 12 feet wide at the base and 7 ft wide at the top, they have140 foot blades. They are about a football field from point to point- they are huge. The blades weigh 225 tons with a 350 ton base and rotate at 15 RPM's and kill 10 slow or unlucky birds a year. California has over 15,000 wind turbines currently. With those in place, natural gas still provides more energy in that state.

Fuel cells show promise, but the cost and storage challenges are extraordinary. Small unit cells are becoming viable for home or small offices. Units cost between $20-$40,000 to build them for 20 to 40 kw. The operating expense is around 3-5 cents per KWH. This is an operating win in most parts of the US. These are wonderful price movements from the early cell creation of $600,000 per KWH during the Apollo project by NASA! General Electric is a large sponsor for fuel cell improvements. GE is making them and making them financially viable. These are commercially deployed on a larger scale with success in higher cost per KWH footprints where the higher capital expense to install is not as apparent to the end user. Look for more large scale fuel cells to help shed load from central generation and support decentralized solutions. Many of these power plants operate in urban environments with natural gas which is reformed in place to extract hydrogen. The reformation takes place in the fuel processing system. A thermal management system distributes the low and high grade heat generated during fuel cell operation. Fuel cells work with water balance and often require no additional water. Once hydrogen and air are flowing into the system, and power is created; the system distributes power loads within the plant itself.

In sum, the green or conservation efforts have not all caught up with the economics of change. More often than not, alternative power is more expensive than existing fossil fuel.

- Hybrid cars are more expensive.
- Regular fuel is almost as expensive as high octane.
- Cloth bags are more expensive than throw away plastic bags.
- 7 cents pKWH for power is less expensive that 9.5 cents for wind.
- 7 cents pKWH for power is less expensive than 29 cents for solar.
- 7 cents pKWH for power is less expensive than 14 cents for fuel cells.

From a generating power point of view, the following represents the existing and alternative power sources.

Source	%of Supply	CO2 Emissions	Comments
Coal	48%	37% of US	Clean coal traps 90%
Natural Gas	22%	20% of US	Cogen option better
Nuclear	19%	1% of US	Leaks, waste, bad guys
Hydro	6%	1%of US	Environmental
Biomass	1%	Unknown	Win for waste
Geothermal	>1%	4% Of Coal	Viable-site specific
Wind	>1%	2%of Coal	Storage issue-viable
Solar-PV	>1%	4% of Coal	Storage issue-viable

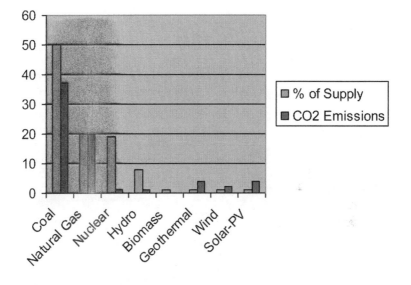

The costs for the various power sources are in flux. The following should be used as guidance for "greater than" or "less than" comparisons for operating expenses without transmission mark-up or expenses. The capital expense and time to market vary dramatically, as well as the impacts of carbon credit trade, government subsidies, bonds, loan guarantees and other incentives.

Coal-	**5.5 cents**
Natural Gas	**6.5 cents**
Wind	**5.5 cents**
Hydro	**2.5 cents**
Nuclear	**3 cents**
Biomass	**6.5 cents**
Geothermal	**4.5 cents**
Solar-PV	**20 cents**

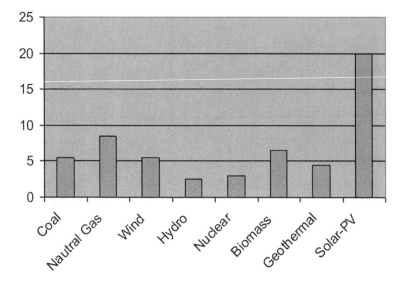

Unfortunately, only 10% of Americans are willing right now to pay more for alternative fuels. This will change. That is the number to change by "choice" rather than at the end of "exhausted resources" and the brutal laws of supply and demand. Protecting our national resources and security, as well as abating global warming will require a 180 degree or systemic change over a relatively short period of time. The use and displacement of carbon is going to get more expensive over time. It will find its way to our national GDP and our disposable income. One estimate has Kyoto compliance at $180 billion annually or .5 percent of global GDP for "hot air." It will hurt, but it will be necessary. For those who were around during the oil embargo and the Carter administration, we know the challenges, both real and imagined. Fossil fuels account for 6% of global GDP. Although energy efficiency has improved 500% over 200 years, it needs to improve in order for us to meet the perfect storm of excess demand and supply shortages

Given our history of lethargic change in the US for these types of challenges, (see war on drugs or war on terror) we have learned we cannot win a war against a noun. A war on energy waste or CO_2 emissions would be the kiss of death at this epic point in time as it relates to our natural resources and national security. Even if the US Government mandated that in the next 4 to 6 years (average vehicle fleet turnover is 8 years) all State and Federal vehicles would be 100% electric or even hybrid, our grid could not support it. We do not have the capacity to plug in that many vehicles at one time. Also, it takes all of 5-7 years to build a coal or gas plant, let alone 13-15 for nuclear, 5-7 for geothermal, and 3-5 years for biomass. We will have to turn this ship slowly and carefully to keep the costs contained and the government's role with the private sector or markets "hand in glove."

Government spending for renewable R and D peaked in the 1970's with the oil embargo at almost $2.5 Billion dollars in 1979. It has stayed between 500 and 900 Million dollars through 2000.

The Department of Energy's solar R and D budget in 2008 is $168 Million.

Venture capital investment in solar energy in 2006 was $164 Million (now being shorted).

The Department of Energy's renewable energy Budget in 2008 is $1.7Billion.

Venture Capital's investment in renewable energy in 2006 was $2.4 Billion.

Federal Subsidies for ethanol in 2006 were $6 Billion.

Federal subsidies for Coal in 2006 were $8 Billion.

Federal Oil and Gas subsidies in 2006 were $39 Billion.

Worldwide Investment in renewable energy in 2007 was $71 Billion.

Federal Subsidies in 2006

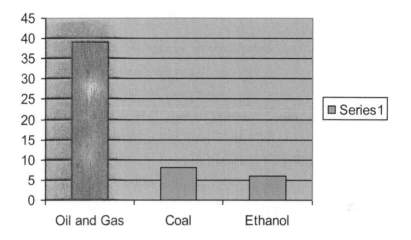

The point is that we cannot rely on market forces or supply and demand opportunities to resolve this. We can not spend our way, or tax our way to energy independence. We are all in this together. For every gallon of oil we do not buy from a State run energy exporter, China, India or other growing economy will step in and buy the availability or excess. In 2007, China bought 80% of the world's concrete because they could, and it was available. India is investing 100 billion dollars in approximately 30 nuclear reactors and is planning to build this over the next 10 years. France and Russia will get a good deal of that work. Common sense prevailing, France creates 75% of their need by nuclear resources, and Russia and India have a favorable and long standing trade history.

Chapter 8:

Financial Impact of the Changing Power Paradigm-Green Gold

Environmental concerns are not unique. The unavoidable energy conservation and energy creation efforts added cost to do business will have unfortunate expenses "creep" into most goods and services in the short term. Some experts have almost a one percent GDP increase in goods and services over the short term as the negative impact of the rising costs in fossil fuels and concentric circles of renewable solutions. This is inclusive of energy crops.

The ebb and flow of the real or imagined financial impact trades like most markets...in expectations. When the price per gallon of oil goes up and this is announced by our 24 hour news channels, we can be sure the price at the pump will go up a few pennies. This happens even though the fuel they were discussing on the nightly news will not hit our shores for 4–8 months following the news segment. These are expectations. They can be translated to food, clothing, schools, staples- pretty much everything.

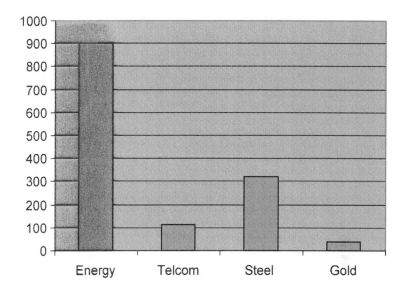

I am old enough to remember the long fuel lines of the 1970's. Meanwhile, gas tankers were photographed in remote parts of the oceans full of a fuel creating the perception of shortage! The cost at the time was on course to go over 100 dollars a barrel and added a 5-7% increase to our gross domestic product. This increase left us at the death's doorstep of economic ruin. It was an economic and high stakes game of "chicken." They blinked, but we suffered and the point was made. The US was helplessly addicted to fuel and the Middle East was the dealer.

There was a push to make smaller, lighter cars. Also carpooling and mass transit became more popular. Solar panels were being fabricated in California, and hence California is known as the birthplace of conservation. This was due in equal parts to our economic supply and the demand needs as well and our ecological concerns.

The world of climate change has morphed from the fringe, esoteric, tree hugging groups to the mainstream economic quagmire. Climate change is not to be ignored globally.

The three main drivers of this consensus and impacting power generation and transmission are:

1) **The Earth is warming.**
2) **Mankind is responsible for most of this.**
3) **Real and negative consequences will impact the Earth in the short term if left unchallenged.**

We start with the power generation as it relates to global warming because power generation is the largest contributor of CO2 emissions. This is growing at a faster velocity of emissions than other sources to be discussed later. The sources of carbon emissions are:

- Electric Power-24.6%: Homes and commercial buildings, chemicals, cement, iron, steel, and other.
- Land use-18.2%: Deforestation, harvest management, reforestation.
- Transportation-13.5%: Road, air, rail shipping.
- Agriculture-13.5%: Soil, Livestock.
- Industry-10.4%: Production.
- Buildings-9%: Buildings, commercial, manufacturing.
- Waste-3.6%: Landfill, waste water.
- Industrial Process-3.5%: Alloys, iron and steel.
- Fugitive Emmissions-3.9%: Coal mining, oil and gas extraction.
- Other- 5%

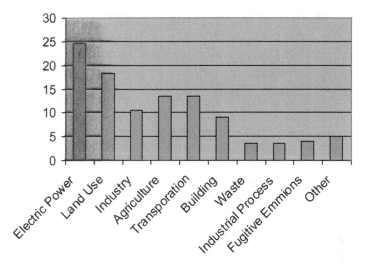

Nationally, the warming of the atmosphere and power generation/transmission is tied at the hip for the largest and most target rich areas for improvement. Years ago, the power generation in the US was almost equally split between hydro, nuclear and fossil fuels. Other, or alternative means of power generation accounted for barely 2% of the total creation or supply in the regulated world.

Reluctantly, power generation has a disproportionately negative impact on global warming via excessive CO_2 emissions. We now create power by almost 70% fossil fuels!

- 50% Coal
- 19% Gas
- 19% Nuclear
- 10% Hydro
- 2% Other

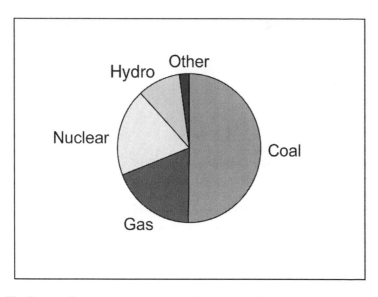

To digress for a moment, according to the Intergovernmental Panel on Climate Change (IPCC). Over the 20th Century, the global surface of the earth temperature increased approximately 33.08 degrees Fahrenheit. In 2005, we had the warmest year on record. There were 1,800 heat related deaths on Chicago, Illinois, and 35,000 deaths in Europe. Since records started being kept 154 years ago, 9 of the past 10 years were the warmest on record! Baghdad had snow for the first time and China had the coldest winter in 50 years.

If you start to incorporate what has been happening in the power generating world, then consider the specific population growth data points and sprinkle a bit of Moore's Law as it relates to technology. Add to this the power demands for technology impact power generation and you will see a straight line relationship between CO2 emissions and the increased demand for power.

So who is responsible? Some of us? All of us? Let's go to the tape. According to the International Energy Agency, the following countries or entities have made their respective CO_2 emissions:

1973 2004 Change

- OECD (30 Countries, USA, UK, Germany, Japan, Spain) 66% 49% -74%

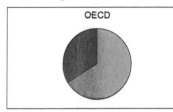

- Former USSR 14% 9% -64%

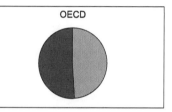

- Asia ex China 3% 9% 300%

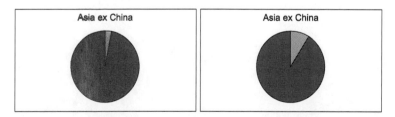

- Middle East 1% 5% 500%

- China 6% 18% 300%

- Latin America 3% 9% 300%

- Africa 2% 3% 30%

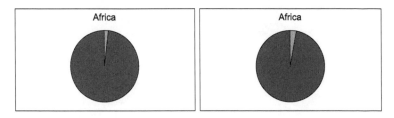

- Others   5% 4% -20%

At a glance, it is fairly easy to see from the power generating data points provided, as well as the CO2 emissions growth or reductions above that the emerging markets of China and Asia are making enormous contributions to CO2 emissions and global warming. The International Energy Agency (IEA) took the time and interest to measure atmospheric greenhouse gas concentrations currently, with given velocities of increase to project what our world may look like for our children.

It is worth mentioning that the unit of measurement for such greenhouse gases are "parts per million by volume" or ppmv. The concentration of ppmv at atmospheric CO2 concentration of greenhouse gas before the "Industrial Revolution" was approximately 280 ppmv. Today, the concentration levels are 380 ppmv with a 33 degree Fahrenheit increase in temperature this century! Projections have the part per million volume over 500 if 2.5 nm particulants are not abated.

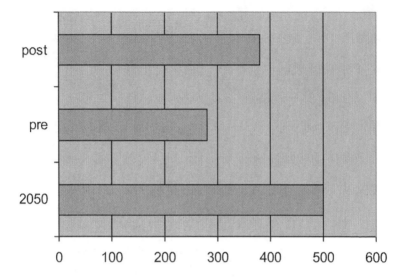

Power generation is responsible 25% of current CO_2 emissions. Industry represents 14% of the total, and transportation creates 14%. These are the target rich areas to conserve or create new low carbon methods of creating and transmitting energy.

Each doubling of the gases can contribute to a 30–35 degree increase in temperature. This would be disastrous by anyone's standards, creating an extraordinary impact on human health, agriculture, potable water and so on.

CO_2 emissions will not stop. For instance, 2-3% of all CO_2 emissions globally come from the coal burning mines that cannot be extinguished. They burn freely into the atmosphere. There is systemic inertia due to the natural ecosystem and anomolic occurrences like coal fires.

With the global contributions of CO2 emissions of greenhouse gases, we are looking at levels of 500 ppmv by 2050 which invites the earth's temperature to rise by 33-40 degrees Fahrenheit for a world population that will have doubled by the same year. In sum, we will be looking at more gases, greater temperatures, less food, less water, and the burden of two times the world's population! This leads us to two assumptions widely held on a bipartisan and global scale:

1) The forecast of climate changes are uncertain. Data points that we have are 154 years old at best. Our forensic abilities to understand the data and geology is outstanding, but the scope and scale of what comes next will largely depend on the seriousness of adaptation of the world's ability to alter generating sources and create other sources.

2) Unchecked, the likely effects of global warming will cause the melting of parts of Greenland and the Arctic Circle. This will add more water molecules to the ocean. The expansion of the water molecules will be added to the heat created by the thermal blanket created by the greenhouse gases. Warmer water creates greater mass, colder water creates smaller mass. The ocean will rise and create coastal shore levels well above today's currently growing at .36 inches per year of 7 inches over 20 years and up to one meter by 2050 by some estimates.

Water Level

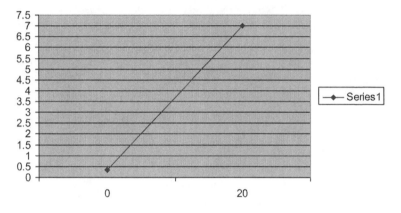

This would be catastrophic for port or coastal cities. New York City, for instance, has 80 bridges and tunnels connecting the 5 boroughs. Most of these have low points located below sea level. Manhattan alone has 20 bridges, and all but two are under sea level. Eighty percent of the top twenty cities in the world are situated below sea level. There are 14 waste facilities at the water's edge in Manhattan. If they go underwater, it would be a catastrophic environmental event. One cannot minimize the concentric circles of risk and damage created by the environmental changes and challenges presented by the increase in greenhouse gases. But to be level headed, jumping in both feet with a doom and gloom, sky is falling scenario, is making everyone an expert from politicians, cable pundits, economists, and others. Common sense required!

Rather than focusing on the catastrophic damage that may come to the tri-state area of the Northeast if a power failure were to occur, let's think about how we can leverage the 80,000,000 gallons of solid waste we New Yorkers "flush" down the toilets. The capture of the methane gas from solid waste is not newly tested, but the urgency has never been greater to pay now for eco friendly and time tested solutions. Shipping waste to other states for their methane use or for fruit fertilization in southern states is not in the short or long term best interest of our urban environments. The change suggested in the power paradigm is a concentrated focus on waste usage and recovery. With most urban environments unable or unwilling to replace pumps, motors, conduit for potable water, sewer and waste treatment facilities, the need is urgent for smart and swift energy conservation practices. The waste treatment and water facilities need to be put in place, as well as energy creation solutions for the natural recourses created by our waste products.

Biomass and bio fuels show extraordinary promise if executed properly. The energy crop currently funded and incentivized have short term negative GDP consequences. What is not known is what is trading on expectations and what is trading on costs or expenses and therefore passed on to the consumer. Switching from fossil fuels to renewables will cost dearly in capital, operation, and carbon credit purchases.

Green Gold refers to the economic boom caused by energy creation from renewable sources and even the negative absorption of power from legitimate conservators or "negawatts." The last two frothey booms we can identify with the energy challenges of today are Y2K and the dot com spending. This wave of spend will dwarf both.

The EU has subsidies for green energy to boost bio fuels to 10% by 2020. That's not just for energy creation by the required distribution to the user.

As the price of energy goes up by over 100% in a year in some parts of the U.S., the appeal of recycled materials becomes more compelling. More than anything, it is a commodities spike. The commodities are waste in its raw form: glass, aluminum, human, and produce.

One ton of recycled aluminum saves $700 of electricity. That alloy specially takes a tremendous amount of energy to create. That is one of the reasons that smelting factories are in Iceland where energy is 3-4 cents per KWH. Iceland, by the way, does not like the CO2 trail that one factory leaves behind.

Methane gas, a product of waste, is not very popular as well. The source is a good use of existing waste, but does contribute to greenhouse gases. In 1988, there were 8,000 operating factories in the U.S. and now there are just over 1,700, according to some reports.

The "not in my backyard crowd" (NIMBY) is being replaced by the "Build absolutely nothing anywhere near anything" (BANANA). A challenge for siting waste energy creation.

The corporate governance and recent governmental or quasi-governmental direction is to convert from the "dump to pump" solutions. We are full of old ideas. Everything old is new again or the more things change, the more they stay the same. Clichés are useful because they are often true.

Chapter 9:

Global Warming: The Cost of Doing Business Moving Forward – Plan Now or Pay Later

Supporters of deregulation of power believe that change will enhance the end user's economic benefits and reliability of the "just in time" network claim that the cost of electricity service will drop due to competition, and eventually force lower prices. At a glance, this appears attractive from a power generating point of view. Cut reserves, operating expenses (people) and increase profitability.

If your region or market has the choice of which generating company to buy power from and we indicated earlier that .5 cents was formerly allocated to regulated reserves that no longer applies. It may make sense for the user to purchase power from the lowest cost provider on the open market. The market for Connecticut may include capacity from Florida Power and Light (FPL). Again, markets are a price and not a place... This is appropriate because the generating company that the user buys power from might have been generated from another physical facility with season or anomolic over-supply, and in this case the benefits can be passed on to the user. There is no apparent difference in quality or known established knowledge of particulant or CO2 emissions or impact. There is no known knowledge if power was bought from a company with the lowest reservation capacity or a company with the worst maintainability or operating efficiency. It's just an "ask" and "bid" transaction. In other words, it is a "blind" buy."

The dirty little secret is that in this deregulated world of no required reserves or operating efficiencies, the "cop on the job" is the market place of supply and demand. There are no replays. There are no yellow flags or red flags for capacity, quality or renewable generating foreknowledge. If milestones established by generating companies are established in S.2191, they will not reach the market place for another 5 years or so. This means we will not realize meaningful net benefits for 10–15 years. Most of the change is realized in the spending and anticipated revenues of "cap and trade." 2018, is a Congressional Budget Office (CBO) milestone for direct spending. Between 2009 and 2018 approximately $1.2 trillion dollars will be spent on enacting the S.2191 legislation, or just over a billion a year. $400 to $500 million of that, or about half, is going into "education and IT" upgrades. That's quite a bit for education and IT, as well as some infrastructure "smart grid" upgrades for a system that already exists and has a human infrastructure.

The likely benefits or consequences of the above are that in the unlikely seasonal or regional outage, supply or equipment failure…we all have deck chairs on the same sinking ship. The users in the same regional footprint will suffer the same consequences of poor planning by one or more generating company in the region (see West Coast blackout and East Cost blackout). The poor planning, aging infrastructure and limited reservation of capacity can created a negative cascading of power. The same shortcomings have contributed to extraordinary price increases passed on to the user of 100% or more in some parts of the Northeast in the past two years. What is the recourse? None. Maybe some protracted and expensive loss of business law suits. Unlike many services; service level agreements (SLA's) are usage or do not exist.

The old antidote of "well if it's that bad, and we ALL go down (lose power), what's the big deal?" Well, the big deal is that most of us work in a global economy and, not a local economy. If my small part of the world has an interruption, my customers and clients out of state, region or country are NOT going to be happy! In this new world economy of "Wikinomics", the virtual value added solutions provider working from home or locally can and does have meaningful real time impacts on the global economic footprint. The user is called the baby boom "echo" user, and there are about 2 billion of them around the globe who make contributions real time all the time. The "Net" generation makes the global economy that much more fluid and reliable, and dependable power that much more necessary. Now, we have a telecommunications system that is capable of meaningful throughput globally, the power to satisfy the central offices, hubs and nodes for the telecommunications system as well as the corporate user has never been more critical. At the same time, the system has never been less supervised. The total cost of a transaction or total cost of ownership (TCO) fundamentals are under unique pressure due to the real cost increases and reliability considerations for energy. When a company, car, telecom, IT manufacturing reaches a point where the cost to transact exceeds the cost to outsource, then the outsource option is more likely. The fables chip company and semi fables car company is here and now. Data centers today are heavy power users and a last bastion of privacy for the financial and post Sarbanes Oxley (SOX), HIPPA and Patriot Act world. The cost for the user to sole source this expensive and intricate data center improvement has money and time values that are better outsourced in many cases to those who do this for a living and can employ cost control measures by scale and best practices, as well as buying power for energy components that have gone up in price two and three fold over the past 3–5 years

The solutions or band aids for the mission critical facilities and many ordinary power users are self help with on site power generator back-ups to take them through any unplanned outages.

Since a vast majority of outages are less than an hour, you may question, "what's the big deal?" The big deal is that the economic and physiological impact that the lights going out creates. Being stuck on the elevator or left in the dark during normal off hour's operations can be traumatic.

As well, in the technology intensive world that we live in, the lost data on the 500 million computers in the US and the millions of servers, routers and storage devices in constant use, the retrieval of data and resetting of proprietary hardware and software by certified vendors can take 4 to 10 times the actual outage duration. It also takes weeks to months more to validate, certify and make Sarbanes Oxley, SEC, NASD, and HIPPA compliant.

Now, there is the developing of important methods of providing the generation and transmission of power that is growing out of the glaring shortcomings of deregulation. Decentralized and renewable sources are coming and it will all be transmitted over direct current (DC) conductors of 768 KV in some parts of the United States!

I have discussed briefly the supply and demand market place pitfalls of creating and transmitting electricity based on the long term and capital intensive requirements essential to keep up with our population growth.

There is now a new class of user who is knowledgeable in the carbon footprint impact of their personal and corporate usage. What is in front of us collectively for energy conservation and energy creation is what former President Clinton called "the biggest economic boom since this country mobilized for World War Two." Right now, we are just at the beginning of this fast paced movement. We are on the cusp of a "Green Job" explosion of activity on many levels. Early movers are often losers, like ethanol, but the successful efforts and business will create over $650 billion dollars worth of investment and hiring over the next ten years according to the Times of London. Bill Gates calls the "Green Economy" the biggest economic opportunity in the US. The defining moment here is not the grass roots "tree huggers" or recycling person. The new driver in the ecological and economic quagmire is the "C" suite corporate governance issuance to reduce consumption of power by 2–7% over two years and start trading inter state CO_2 credits to offset personal and corporate fingerprints of usurious carbon usage.

When a person or corporate user realizes that it takes 966 tons of coal to run a computer of 100 watts for a year... then they realize that 70% of burnt coal and most fossil fuels is lost in heat (the waste product of energy) 70% of the net energy is lost in transmission of energy via HV transmission lines to LV distribution lines, through step down transformers and rectification from AC to DC to AC. The net or end result of the burnt fuel is 10% of that 966 pounds! Bio fuels have lower calorie value by almost have. This bothers people. This bothers not just engineers or scientists, but legislators and policy makers. This also bothers home makers and children, and small business owners and laborers. This touches and makes sense to all of us.

A whole new method of power creation and distribution is in the process of being deployed. It is decentralized by being built closer to the need and has a lower carbon footprint. The deregulated companies sometimes embrace the load shedding from their over stressed networks, and sometimes they make decentralized Independent Power Providers (IPP's) less attractive due to the economic cannibalization of their revenue streams. This is truly a "region by region" paradigm with various legislative consequences impacting the most extraordinary "just in time" network in the world. Look for involuntary load shedding by the utilities to the home and businesses. TXU is currently employing this practice. They can remotely turn down air conditioning by the "Zigbee-Enabled" demand response by "i-thermostats." This is appropriate for utility or resources abuses like people who water their lawns during droughts but it also smells of "Big Brother." More to follow on this one. Stay tuned.

The economic costs to the world due to the dysfunction of the regulated and deregulated power creation are staggering. Then you add the cost of transmission, as well as global warming contributing up to 25% to irresponsible power generation and aging transmission. The first impact of energy concerns is demonstrated by conservation, the second impact is the multiple centralized and decentralized energy creation and the last impact will be meaningful regulation. Parts of the Clean Air and Clear Sky's Act as well as the Energy Security Act of 2007 will extend to meet the market of appropriate energy conservation and creation methods. Some estimates of the pass through cost to turn the tide of ineffectiveness through conservation and alternative power creation are between 1–3% of the respective Gross Domestic Product (GDP) annually of the country. The global temperature of the earth could reach to 3–30 degrees Fahrenheit by 2050.

If the CO2 emissions levels reach 550 ppmv, which is almost twice that of the pre industrial age ....we have a problem. Remember, we are at 380 ppmv today and it took us 100 years to double in population.

The cost to limit the CO2 emissions, not just through generating of power but conservation and creation methods for industry, transportation, buildings, agriculture waste, land use etc. vary between 1%–3% GDP for net benefits or savings (yeah!) or up to 5% cost to mask dysfunction (boo!).

We are currently in the embryonic or stage of creating methods to reward conservation and energy creation methods that meet exacting measurable milestones or

expressions. The term "Green" is unclear. The search engine Google has over 425 million hits off the term "Green." This "Green" component will alter the math we use to measure a company's performance. There will be bottom lines, below the line and a "green line" for the inclusive value of the carbon benefits accrued to a company for efforts to be environmentally responsible and the costs to support the same. The challenge to date has been one that cannot reward and cannot be measured. The data base, or data points for creating incentives for utilities, car makers, building owners, industry and governments to do better is guess work at best right now. That is why no meaningful legislation or incentive program exists currently in the United States. Republicans and Democrats alike do not have enough empirical data to craft informed legislation. Now, we can get upset about how long it took to embrace the gaseous and warming realities, but unfortunately we are where we are.

The push is to measure existing conditions, apply new technology, and reduce the emissions by conservation and alternative power creation inspired by the best practices. The average American uses approximately six times the average per capita energy world wide and 50% more energy than the average European. It is easy to say that America is a target rich environment for both energy conservation and energy creation.

Sectors particularly likely to be affected in the short term by the shortage of power, interruption of power and high cost of power are:

- Utilities-Generating Companies
- Utilities-Transmission Companies

- Petroleum Products Companies
- Mining Companies
- Metals Companies
- Insurance Companies
- Clothing Companies
- Building (materials) Companies
- Pharmaceutical Companies
- Construction Companies
- Real Estate Companies

Companies that will prosper or flourish in the energy strapped and climate changed world will be the forward thinking companies that are looking to "self help" solutions for energy conservation and creation by thinking outside the grid.

Solutions and mandates start from the top down or from the "C" suite to the facilities group. Much like the new need for a "business continuity" expert to be on site (not a part time assignment of risk management or HR), an energy conservation or energy creation silo needs to be established with a clear mission, resources, intellectual capital and human infrastructure.

Once these disciplines are created to reduce emissions, and work more efficiently we can collectively establish the Global Emissions trading systems intra country and inter country by company, utility and country. The size and credits currently are large enough to create markets and are going to grow at a coefficient of ten or greater in the next 3–5 years, judging by the emissions growth in India, China and other emerging markets with little or no restrictive covenants for emissions.

In China, the strip mining and surface scars of coal mining are exacting their toll as well. Unnatural amounts of minerals are flowing into the rivers and causing massive sickness for river communities and cities. Cancer in these communities is 5 to 10 times greater than the regional rate, with a 2 dollar a month per person compensation for the contamination and health care!

16 of the top 20 cities on the world are in China, and unfortunately most of them are in flood zones. If the rising tides do appear in the next 20–40 years, there will be multiple and extraordinary migrations to higher land. China's expanding economy is remarkably successful and extraordinary to watch. It is not dissimilar to the US's economic expansion during the industrial revolution. Money first, environment last. Europe is no better. We learned our lessons sooner and bottomed out. We are on a collective recovery program now. The first step is recognizing that you have a problem and your life (country) has become unmanageable.

Where there is adversity, there is opportunity. Combined Heat and Power (CHP) or tri generation is an old solution to a new problem. Cogeneration creates steam to turn turbines and is being modified from gas only with a 20–40% power to "Gas to Liquid" (GTL). This turns gas (of limited supply) with coal, or biomass at source to liquid fuel by blending it with pure oxygen under heat and pressure to produce synthetic gas, which in turn transforms into a diesel-like molecule. GTL will likely require cooperation with energy and gas companies.

New and lighter materials will be a by product of need. Storage media to support the photovoltaic cells or panels will make solar power more commercially viable. Lighter materials will allow cars to go farther. I'm sorry, but I do not see SUV's or 8-9 seat cars getting 35 miles per gallon unless they are made of some light material- like cardboard. Some of this is just basic physics. Most cars are simply too heavy to get good mileage. With safety restrictions for crashes and minimum support for 4–5 seat cars...then throw in a few 400 pound batteries so you can idle when needed, you are not getting these cars to get great mileage without a significant speed, distance, weight (safety), size and comfort concession. New alloys are required or 180 degree changes in design and driving habits need to be implemented. This is highly unlikely due to the redesign expense of the drive train modification costs that we are currently using. The greatest opportunity for improvement cyclically is when a car is replaced every 4–6 years. Historic drive train changes yield under 2% benefits to weight. Not the rich inventory we were looking, for but better than nothing. More surgery will be required to whittle the weight down and create a significant redesign. Look for smaller, higher cars, no magic here.

In 25% of carbon creation, the fossil fuel burning process, clear coal production is likely to add to the cost of energy and reduce energy output, but it is perhaps worth it. We are currently and partially legislating our way through this financial and ecological minefield of greenhouse gases and CO_2 emissions. There is no easy answer on what to do with the carbon, or the expense of the carbon capture process.

The reserved fuels for R and D and early stage development for alternative fuels are remarkably disproportionate to the financial and ecological consequences if half of what is forecast actually happens.

The American initiatives are well behind the European initiatives. The ECX (European Carbon Exchange) dwarfs the CCX (Chicago Carbon Exchange) in volume, liquidity, law, governance and consequences.

Energy conservation and utilization are more commercially accepted with lifestyle and economic data points available to measure against. Cogeneration is a common source of energy for hotels, state of local government facilities in Europe. Cogeneration captures the heat from generating energy and creates hot water for showers, restaurants, perimeter heating, and so on.

Open floor plans require by code a significant percentage of natural daylight to penetrate the floor directly. This open flow concept means fewer private perimeter offices. The open plan creates a more cheerful atmosphere which translates to better mental health. Also, the open plan lowers energy bills due to the natural sun light reaching deeper into the floor and illuminating more of the office space. In the U.S., we value greatly the power and prestige of the windowed office. Also, corner offices are better than just perimeter... even more power and prestige!

Europe is not perfect by any means, but they have done more with CO_2 reduction efforts, as well as energy creation. They started sooner due largely to the fact that they had far fewer natural resources than we have had. The US was resource rich, but we spent them freely, until the waste and consequences became unavoidable.

Our collective responsibility is getting China, India and other parts of Asia to become leaders in controlling their own gaseous emissions.

Asia is going through their own industrial revolution. Financial models or incentives need to offset the adverse effects of growth and waste. A CO2 emission exchange market in China would be a great benefit to the world at large. An inter country Carbon Credit Market could be created where one could off-set the damage done financially with benefits created elsewhere. The one thing I can say with the utmost confidence is that it is very likely in the short term that the cost to improve energy generation and effective transmission or distribution will cause an increase in prices. R and D efforts will be amortized with real improvements to meet the market. The public sector will legislate to support the private sector improvements which will determine the demand curve. Overall reliability will be less than our recent experience due to severe weather swings, include more severe weather. It is getting harder to tell the difference between severe and catastrophic weather with cable TV pundits now experts. Have you noticed the collective disappointment when there is no or a low death count?

Only through an extraordinary effort can this story have a happy short term or long term solution. Partisan sound bites do help the awareness, but do not bring us closer to long term solutions inclusive of public sector and private sector cooperation. This is not "fire and brim stone stuff"... cats and dogs living together." What we do not want is a future of scarce resources and expensive alternatives affecting our lives.

Chapter 10:

The Government's Role (Leading or Following)

As of 2008, the Environmental Protection Agency mandated that individual states can not set mandates for the Federal Government to create deadlines or specifications for auto or energy creation guidelines. I take this as a step backwards for our collective interests. We, as a country have learned by way of example from our states and their successes and failures in various initiatives, especially environmental initiatives.

The Clean Air Act of 2002, The Clear Skies Amendment of 2006 and the 2007 Energy Independence Act are wordy documents with very little specification on how we are to actually implement air quality and water quality initiatives in both the short and long term. They are a start.

In the Government's defense, they can't fix what they cannot measure. "Energy Star" is the current instrument the Government now employs to measure power demand, quality and conservation measures. The US Government data centers have taken up approximately 10% of the total US Government's total demand between years 2000-2006. The Government's Department of Defense (DOD) is leading by example in energy conservation. Of the 78% of the Pentagon's total government's power consumption, almost 12% of the DOD's energy consumption is from renewable sources, primarily wind. The Air Force alone accounts for 40% of the entire federal government's renewable energy usage. Geothermal power heats hangers with Aircraft's unique alloys requiring a warm ambient temperature. Better than electric! This is perhaps an aggressive position due to jet fuel emissions and recent years of overspending.

I believe this was their first hand reaction when confronting tree huggers, industrialists, and legislators. There is no shortage of passion from any one group. I would only encourage the data point collection and manipulation with the Government as a client and reference point for ordinary and mission critical power loads , quality, and velocity of growth modeling.

- The tree huggers want to save the planet for future generations. This is a noble and virtuous cause for our quality of life as well as for future generations.
- The industrialists need and want to do business at the speed of the market place in a cost effective manner. Any deviation in energy reduction or conservation creates a real or perceived increase in the cost of doing business. Some economists have the short term increase in the GDP at 1–3% to comply with existing or proposed legislation.
- The legislators want to represent business interest (campaign contributions) and the general public's interests, which are generally not always on the same page. In the 110th Congress, with the development

of Senate Bills S1419 and S.2191, we have new energy laws which 65 senators voted for and 27 senators voted against, 7 senators abstained. The states with some of the worst air quality voted against Texas, Louisiana, and Georgia. Also, the oil and coal dependent states like Oklahoma, Mississippi and Kentucky also voted against the bills. North and South Carolina, I just do not get. They both have air quality issues, as well as water issues. Residents are often in conflict with infrastructure improvements of generating or transmission companies needs.

Several environmental entities and initiatives have been created since the 1970's including:

1. Richard Nixon's response to the OPEC oil Embargo to be energy independence. (This goes back a while)
2. National speed limit of 55 mph to conserve gasoline.
3. Gerald Ford moves back energy independence to 1985.
4. Fuel standards set at 27.5 mph.
5. Jimmy Carter stated energy independence is equal to war and creates the Department of Energy.
6. Iranian Revolution contributes to oil doubling in price.
7. Carter commits to energy independence by 1990, and puts 32 solar panels at the White House.
8. Ronald Reagan stops solar program in 1981.
9. Ronald Reagan removes solar panels from White House in 1986.

10. George Bush reveals plan to reduce dependence on oil 4 days before the US invades Kuwait in 1991.
11. Bill Clinton proposes 40 mph cars by the year 2000. (Not)
12. 55 mph limit is repealed in 1995. (And we're off!)
13. Kyoto Protocol signed by 174 countries. (A start, look for "Kyoto 2")
14. 2000–2001 California hit by rolling blackouts.
15. 2005, US Congress triple the amount of ethanol to be mixed with gas by 2012.
16. George W Bush vows to cut Middle East oil imports by 75% by 2025.
17. 2007- Congress raises mph for commercial traffic to 35 by 2020.
18. U.S. Will not allow states to supersede Bill s.2191 parameters.

Thank you Mother Jones.

The more recent legislation has received most of the current attention due to:

- The public's non-partisan view of global warming.
- Global economies' insatiable appetite for power in newer economies such as China and India (Chindia).
- The United State's awareness of its global contribution of the carbon footprint. For example, if Texas were a county, it would be the 6th largest CO_2 contributor in the world.
- Moore's Law of data bandwidth is doubling every 18 months and the impact this has on our computing, data storage, data use and PDA's energy needs (now at 3.5% of total US consumption).

The Congress, Senate and the Environmental Protection Agency were taking action in 2007 to amend the Federal Implementation Plan's (FIP's) for the Clean Air Interstate Rule (CAIR) to provide for automatic withdrawal of the EPA's approval of a full state implementation plan. All of these states are to reverse their plans will include control measures to reduce the emission of Nitrogen-Oxides (NOX) and Sulfur Oxides (SOX). What was emotional about this legislation was that the state of California had emissions rules and criteria that were more aggressive than the EPA's! Historically, California does lead the way with green energy conservation practices and negative consequences of energy conservation and impacts of energy deregulation.

The final rule date was effective January 16, 2008, and the collective legislation is meant to address the ecological and commercial impacts or realities that:

- A good portion of the existing energy creation infrastructure in the US over the next 20–40 years will be inefficient or "out of service." The useful life of the improvements in many cases will have expired.
- Approximately 90% of the world's "current" oil supply comes from "state run" firms. They effectively renegotiated the multinationals out of their footprint. They are taking the fixed assets, R and D as well as intellectual capital to find the resources. Multinationals are reduced to taking minority positions in their own assets and have to grin and bear it! Fifteen of the top 25 top oil companies are State owned according to "Petroleum Weekly."
- Currently, CO_2 levels are at 380 parts per million volumes (ppmv)—up from the pre industrial

revolution 280 ppmv. These are the highest levels in the last 750,000 years.

- Global warming and rising sea levels impact of 7–30 degrees Fahrenheit over the next 80–100 years. We can expect this if warming is not abated.
- Global sea levels will increase .36 inches a year, or 7-12 inches over the next 20 years.
- Global shortages of potable water or drought conditions will continue to shut down nuclear facilities in the US and Europe due to minimum water levels needed to create energy and cool reactors.

There are a host of new or pending laws in various stages of completion in reaction to an over reliance on foreign petroleum products, limited fossil fuels and global warming. As well, we are seeing in the US and Europe subsidies or tax incentives to promote the creation, storage, sales and usage of clean fossil fuel and alternative energy. More incentives are required to expedite timely creation and distribution of alternative centralized and decentralized power sources to relieve the nation's grid generation reserves, capacity and distribution.

The new and relevant legislation that we will be responding is the Energy or Climate Security Act of 2007. The highlights are:

- Cars must reach 35 mpg by 2020 (major surgery required for most cars).
- $450 Million will be allocated towards alternative vehicles with $25 Billion in direct loads for automakers.

- $500 Million in grants from 2008–2015 for biofuels.
- $50 Million in grants for solar powered workforce training.
- $1.2 Billion to develop carbon sequestration (Yucca Mountain cost $11 Billion).
- Establish Office of Climate Change.
- $21 Billion in tax breaks and incentives for gas and oil companies.
- By 2020, utilities must provide a minimum of 15% in renewable energy.

The Congressional Budget Office, or CBO has reviewed and summarized the Act. They indicate that the newly formed Climate Change Credit Corporation will implement the allowance the EPA sets forth to specific users for buying and selling. S.2191 would require the EPA to establish two specific cap and trade programs for two distinct climate killers. They are GHG's or green house gases (CO2, methane, nitrous oxide, sulfur, hexafluoride and perfluorocarbons) and HFC's or Group 2 GHG's. The "Corporation" will "auction" allowances to emit one ton of carbon dioxide or have one ton of entities to buy and sell allowances among themselves. A portion of a company's compliance would be created by an EPA certified company verifying an "off set" of activities that reduce GHG emissions or the capture or sequestration of GHG's.

According to the Bill, Federal revenues would largely depend on the value of allowances created by the Bill. In effect, they are making the market for revenues. Penalties for non-compliance and fees collected to administer the legislation would add a varied amount to the revenue base.

The penalty should be a hammer. The estimated emission allowance for Group 1 GHG's is about 400% more than the market can bear in the US at this point. The Chicago Carbon Exchange values the credits at 6 dollars per ton right now. In the EU, it is currently 25 Euro, or $37 dollars per ton. The US market is voluntary and the EU market is mandatory. Until the Kyoto Protocol or other Inter Country mechanism is employed, there is no way to trade credits, usage or sequestration.

Kyoto was crafted in 1995–1997 with milestones in 2010. How good or relevant can that data be? How much should we spend to slow warming? At what point does it make more sense to pay for damage rather then slow emission particulates? Currently, we loose 200,000 people globally to heat related deaths. Currently, we loose 1.2 million people to cold. See where I am going with this? $7 to 15 Trillion dollars to pay for "hot air" and your benefit 5-7 years of emissions.

The finances for the emissions program are established into seven funds for the years 2009–2018 The following is a summary:

1. The Energy Assistance Fund-$64 billion. Low income assistance.
2. The Climate Change Worker Training-$12 billion. Training for workers included.
3. Adaptation Fund-$31 billion. Education for fish, wildlife, and climate change.
4. The Climate Change and National Security Fund-$16 billion. Would finance recommendations for climate change.

5. The Bureau of Land Management Emergency. Firefighting-$2 billion. Supports land management and wild lands.
6. The Forest Service Emergency Firefighting Fund-$6 billion. Fire suppression on Federal conservation lands.
7. The Emergency Independence Acceleration Fund-$6 billion. Support research by DOE.

The goal is to reduce the GHG's by 2050. This is a noble and appropriate goal. I trust the writers are sincere in their efforts, but are misguided in their assumptions and math. I believe there will be amendments to the Act, and future legislation as more discovery and intelligence become available. For instance, the Act allocates $3.7 billion over 9 years, or $370 million dollars per annum "to be used to support EPA personnel, contractors and information technology necessary to implement the legislation." Now, I realize that super sized funds may be required to solve this enormous challenge, but these are funds with no name as of yet.

There is now a national movement, both grass roots and institutional, to help make alternative and renewable energy more competitive over the near future. The state movements are currently being trumped by the EPA for leadership in this area, but that is likely to change.

Recent actions promoting these changes are:

1) The Energy Independence and Security Act of 2007. This was intended to move the United States closer to energy independence and security, and

increase the production of clean renewable fuels to protect consumers and businesses. This was also intended and to increase efficiencies of products, buildings and energy production and distribution. This was enacted as well promote carbon and greenhouse gas capture and storage.

2) May 2007: The US Senate held hearings to significantly increase motor vehicle fuel efficiency standards.

3) September 2000- Governor Schwarzenegger signed assembly Bill 22 which would require car emissions to be reduced to 80% below 1990 standards.

4) Bush administration set "goals" of reducing oil imports from the Middle East by 75% by 2025 and to increase Clear Air energy research by 22%.

5) The 2005 Energy Legislation doubled the renewable energy budget to $852 million for 2009.

6) 2002- Clean Sky's Amendment attached $1.2 billion towards alternative fuels for transportation.

7) Kyoto Protocol- covers the EU and 35 other countries to reduce greenhouse gases by 5.2% from 1990 levels through a cap and trade system. The US and Australia did not ratify the agreement.

8) EU- set a goal to generate 10% of total power from renewable sources by 2010. Germany is decommissioning their nuclear facilities by 2022.

9) Renewable Portfolio Standards (RPS) - Requires electricity providers within states to obtain minimum percentages of their power from renewable energy resources. Currently, 20 states including the District of Columbia have RPS policies in place.

10) The Chicago Futures Exchange/Climate Exchange-voluntary but legally binding program for reducing and trading greenhouse gas emissions. Members must reduce greenhouse gases by 4% below baseline 1998-2001 and by 6% by 2010!

11) European Climate Exchange- program for EU trading of credits and use of greenhouse gases. Currently, at 25 Euros a ton per year or $2-6 dollars per ton. The US voluntary market is 6$ per ton of carbon! (Remember, you are buying hot air.)

12) World Business Council on Sustainable Business-The WBCSB is an organization of 180 international companies, with a commitment to sustainable development through economic growth of ecological and social progress.

13) The Pew Center on Climate Change. The center brings together business leaders to make policy and bring together non partisan policy makers for global climate change. This was founded in 1998.

14) Carbon Disclosure Project-Started in 2000. The CDP requests information of the FT 500 for risks and best practices as they relate to large operations regarding climate change. Recently, 48% of FT 500 indicated that climate change is a risk to their commercial positions, and have implemented a greenhouse gas reductions program.

We continue to play "catch up" globally with the energy conservation and energy creation dilemma we are collectively in. Some parts of the world are performing better than others. Sometimes change is inspired from the bottom up, and other times from the top down. I would like to emphasize change from outside as the most prevalent today. Because of technology changes and expanded IT band width with the PC and PDA, changes in business and economies are happening real time with collaborative efforts and relative ease. The new model of innovation as it relates to real time improvements and the sharing of intellectual capital is called "peer production." Low cost software and collaborative "salons" are the place where the best and the brightest meet and discuss and criticize in order to improve the status quo. This virtual epicenter is the way of the future. Those who draw the lines of isolation will find themselves falling back in progress and playing market place "catch up." The "bricks and mortar" paradigm is not dead, but dying. Cooperation and sharing of ideas and experiences is required for US companies to succeed. Some call this "coopetition." In the US, it appears we have a history of being reactive rather than proactive as it relates to changes in the environment and how they impact on our capitalistic society. Do not get me wrong, I believe in the market system and capitalism. I believe we should be compensated for services and value added solutions, but the markets have not been efficient in responding to the environments.

The planning required to build coal, natural gas, and major hydro and nuclear facilities is extraordinary. Our nuclear facilities would not be built at all without government help. In 1960, a new nuclear facility cost approximately $70 million dollars to design, build and commission. In 2008, that number is approximately 4 to 6 Billion dollars! These are the highest cost single private investments in the world. The nuclear lobbyists were looking for approximately $50 Billion to guarantee loans for private companies, and only got $18.5 and $6 Billion in tax credits. This is hardly enough to meet the demands we are facing with the uncertainty of the decommissioning program expected with our existing nuclear facilities.

Nuclear power is plagued with environmental and security concerns. Some of these concerns are more legitimate than others. Sixteen percent of the world's energy generation comes from nuclear power, that is equivalent to hydro.... both have challenges. Ten years following Enrico Fermi's first atomic fission chain reaction, President Eisenhower and our government were faced with what to do with this new man made creation. This was our government's science project and viewed at the time as "Atoms of Peace" for energy deprived regions of the world. Westinghouse built the Shippingport Power Station in Pennsylvania to feed Pittsburgh from a 60 megawatt station. This is considered small by today's standards. Then, the power was "too cheap to meter" and "too expensive to build." Construction and maintenance expenses in the government regulated world of power made it impractical at the time. When you consider the Three Mile Island event of 1979, which was almost catastrophic to the point of explosion and evacuation, we have not permitted as a government a new facility until 2005. President Bush celebrated the 2005 Energy policy, which set the stage for the 3 sites under construction starting now. "Without these plants, we would release 700 million metric tons more of carbon dioxide in the air each year."

I am not sure that is correct, but the point is clear. Nuclear power has a new place in the power generating solution matrix. It is far more cost effective than wind. This is a surprise to most people. A Canadian study shows that to create the same energy profile, 1031 wind turbines would be needed to create the same power as just one nuclear reactor (CANDU 600) or 3 Billion vs 1.4 Billion. The wind scenario would require approximately 650 square kilometers. There are uranium costs and maintenance costs for both energies as well as different levels of reliability. The cost per KWH of coal and gas fires creation has more than doubled in the past 4 years, and has far less carbon impact. The bad news is that while our government was away from the technology in the regulated world, the French led the way in construction of nuclear energy. From a best practices and national security point of view, we need to emulate them. During the 1970's oil embargo/crisis, they shifted dramatically to nuclear generation. France invested approximately $160 billion dollars in their programs at this time. This also makes them the largest exporter of nuclear power in the EU. Fifteen to Seventeen percent of the world's nuclear power comes from France and 6 of the 20 applications expected to be submitted to the Nuclear Regulatory Committee prior to 2010 are from the French nuclear company, Areve.

But enough about nuclear energy. The push by our current administration and the mood of the people appears to be supportive of strategic planning to conserve power. We need to create alternative methods of energy swiftly and with low environmental impacts. The prevalent opinion of the people is driving the strategic and now tactical reactions of the deregulated power companies. The government is unabated. What was once viewed as subjective and silly, is now viewed with data points and dollars.

Our reactive and collective opinion is best memorialized in the Energy Independence and Security Act of 2007. Minus a few exclusions that were meant to soften the blow to the multinational oil companies and our own oil intensive industry, it is a good first pass. It addresses the emotional buttons that are being pressed every night as we watch speculators drive fuel prices up by over 100% in the past 12 months. Many forms of landmark legislation like The Patriot Act, Sarbanes Oxley and the HIPPA requirements, look for future amendments to provide more specific guidance or "teeth." The Bill originally sought to cut subsidies to the petroleum industry in order to promote oil independence and promote alternative energy. That would have been too brutal a blow to the US auto industry. We will take baby steps for now and focus on energy conservation measures, monitoring and power creation. These should be baby steps. The big bets we make can be catastrophic (see ethanol). You can see the obvious need for change. The information I have gathered by witnessing the ethanol subsidy mess the government has gotten us into and the concentric circles of cost and inconvenience the food shortages have provided is that until some of these technologies have been shaken out and the goals or intentions of the lobbyist are truly known, we may be fixing a "fix" for some time to come.

The key goals of the Energy Independence and Security Act of 2007 were to:

- Improve vehicle fuel economy—automakers are required to make cars capable of 35 mpg by 2020, provide incentives for electric or hybrid cars and have federal auto fleet meet or beat new standards.

- Increase production of Biofuels from 5 billion gallons in 2007 to 36 billion gallons by 2022. 21 billion gallons need to come from "non-corn" based sources.
- Improve energy standards for appliances and lighting. Incandescent bulbs to be banned by 2014.
- New initiatives and incentives for more efficient building operations, as well as manufacturing.
- Requires all Federal building to use "Energy Star" products or those viewed by the government to be energy efficient.
- New and renovated Federal buildings must reduce fossil fuel use by 55% from 2003 levels by 2010, and 80% by 2020, and be "totally carbon neutral" by 2030.
- Accelerate R and D for renewable energy technologies.
- Create "Green" job training.
- Create an Office of Climate change within the Department.
- Small business energy programs can provide loans for energy efficiency.
- Create "Smart Grids" to modernize electricity grid and improve reliability and efficiency.

If most of our collective $CO2$ emissions and six particulants were from power generation, eclipsing transportation, building and manufacturing, wouldn't that be a good place to start for meaningful change? The copy I provided above summarizing the Energy Act is proportionate to the copy of the act. There are plenty of emissions and conserves of power quality, supply reserves and reliability concerns, but very little copy. Smart Power Grid sounds interesting, but the legislation has not addressed the challenges created by the deregulation of power in America, and what that has done to our national security and quality of life. The data points being collected have not been vetted and are questionable, as voluntary intelligence can be unreliable.

In the newly formed deregulated companies defense, their margins are thin for creation and distribution. Carbon tax credits and the exchange of same need to be implemented nationally and globally as swiftly as possible so there is a place to offset the extraordinary capital expense required to make the newly deregulated system both reliable and eco-friendly.

Without Government bonds, loans, tax incentives or pilot programs, these facilities will not be built due to these extraordinary expenses and the duration or time value of capital needed to build. Private or decentralized solutions can not and will not be mainstream until the user is met legislatively with open arms by the regional transmissions-1,000 power companies and 600 generating companies, who often have the home court advantage in tariff, cost per KWH, capital expenses and operating expenses in these locations.

If the alternative energy sources are under 2% of current generation and the decommissioning of nuclear and legacy facilities is inevitable, we are collectedly looking at a negative absorption of energy creation. That means:

1) We continue to need or demand power 2–3% per annual for ordinary purposes and 10% of that collective demand is growing at 20–25% per annun for mission critical or data center usage.
2) We will take off line the most inefficient or "dirty" generating facilities, or will continue to deny permits for new facilities in regions of the US and world that have poor air quality (parts of Georgia, Texas, New Jersey, New York, California, ect).
3) Require in long term and long lead time solutions paid for or subsidized by the government like nuclear, tidal, bio waste, jatropha, or other biomass solutions.

Plan "B" for the above challenges will come in the form of conforming to 'caps' for CO_2 and greenhouse gas emission power plants. The acceleration of retrofitting existing fossil fuel burning plants is required to take these "Oh my God" (OMG) component out of the next 20 year supply and demand curves for energy.

We are looking at the perfect storm of increased growth and limited supply. Conservation efforts, if excluded perfectly will not take us away from the OMG factor. Conservation is good, but clearly more emotional than empirical in terms of long term energy solutions.

Short term, and I mean now, carbon capture technology will need to be employed for retrofits on existing fossil fuel plants and the design of future plants. The bad news is that the process of carbon capture is expensive and the process reduces the "net output" of power plants.

In short, we will get less power that costs more from fossil fuels (coal, coke, gas and liquid). The good news is that we will contribute far less to the destructive CO_2 particulates and other earth warming greenhouse gases and build power plants under 5 years to meet demands.

Look for more on mandatory carbon sequestration for new and existing power plants. The most recent legislation to meet these goals is represented in the "Moratorium or uncontrollable Power Plants Act of 2008" was introduced in Spring 2008. The spirit of the Bill is to capture, permanently sequestrate and dispose of up to 85% of CO_2's emissions. So as not to dismantle or catastrophically interrupt our economy, the ramp in or time to comply will be placed in over time, but long term; look for power from fossil plants to be more expensive, less efficient but more available.

From an "Energy Darwinism" point of view, this inevitably will place fossil fuels towards the bottom of the long term energy generation "fuel chain."

Chapter 11:

Acts of God, Human Intervention, and Power-A Perfect Storm Forming

Based on current and useful intelligence, global warming's impact of new and excessive CO_2 emissions is now an everyday science. There will continue to be holdouts like those who believe man never landed on the moon, the events of September 11, 2001 was a U.S. conspiracy, and OJ Simpson was innocent. It is not good. But how bad is it?

The grown up and non-partisan view of the environment is that greenhouse gases and 2.5 micron particulants are negatively impacting our environment with increases in velocity of density. On this, we can agree to disagree. What will the world's temperature be in 2040? Will the agricultural, health, economic, and social impact be manageable? Will the world's population double by that time? Only time will tell.

What we have are relevant data points that lead us to some strategic and alarming conclusions. Energy is similar to the world of technology, where the speed of needs far out paces our speed of solutions or augmentation. What we rarely budget, and really can't budget is what we don't know! By that, I mean there will be both strategic and tactical good and bad news on the horizon to be embraced and incorporated in the problem and solution side mission critical facilities, as well as global in the problem solution. We just can't tell what they are yet. This makes scalable and burstable solutions to day one needs for power users that much more challenging.

The financial exchanges are swiftly merging and consolidating as the volumes of trades are decelerating into other financial vehicles highly leveraged in nature. There are more mutual funds than stocks. Soon, there will be more hedge funds than mutual funds due largely to the fact that hedge funds can take both sides of a trade, long and short. Hell, there are markets for the weather... literally. You can actually bet on the weather! I thought we reached a collective new market low when we could make bets or buy stock on how a movie was to do (sell) over an opening weekend...but the weather!

Carbon exchanges will trade the same way. Add a little more risk due to uncertain energy sources, weather conditions and county specific legislative guidelines and the value of the "carbon credits" or permits could far outweigh the value of an expiring fossil fuel with finite supplies.

The "Acts of God" impact on the grid has much to do with the data points we have collected over the past 154 years of memorializing them. The data on temperatures, snowfall, high winds and water levels as well as more recent data on cost or replacement damage and injury/deaths helped regulate and now deregulate power companies plans for the seasonal or anomolic events impacting the grid.

We must understand that early data points were likely written in crayon and memorialized by people who were not really even involved in a science but rather record keeping. We need to look at early data points with a healthy skepticism. For instance, our 100 year storm mapping is based on only a 10 year (a minimum) or automatically downloaded data from water level monitoring sent to Geostationary Earth Orbit (GEO's). That means 10 years of data memorialize what we call "100 year" events. The same with the 500 year flood planes. Unless we are tracking Noah's Ark, the 500 year plane is a guess of what it would take to fill the earth to raise the water level to a certain level. This does not take into account new levees, dams, irrigation, or the movement of the earth.

What we do know is that the earth is warming, coastal water levels are rising, and port cities in low lying regions of the world are at risk of storm surge and coastal water level "creep." These frequent, but stronger storms are occurring due largely in fact to the warmer ocean and ambient air temperatures.

The severe weather of storm, flooding, and drought are difficult to plan for on a global basis. They truly have local impacts on the global economy or "Gloconomy." If we lose NYC, London, Frankfurt or Shanghai due to regional Acts of God or human intervention, the real and physiological damage to repair and restore infrastructure as well as confidence is a bit overwhelming the local and global levels.

By way of example, recent consequences to severe Acts of God have regional impact. Mission critical facilities generally should not have primary, back up or disaster recovery facilities within a minimum of 30 Euclidean miles of each other. The theory is that the Act of God impacting one power network, substation or transmission line will likely impact facilities within a 30–250 mile radius. Specific Acts of God sensitivities like hurricanes, tornadoes, earthquakes, droughts, and floods have greater distance sensitivity. In general, Acts of God, predictable weather, and anomolic consequences of human intervention, such as terrorism, need to be considered.

St Louis, Missouri, had a very severe weather history in 2006, according to the National Weather Service. The impact on power to the Customers, both residential and commercial was catastrophic. The multiple storms of high winds and rain affected the utility Ameren with costs of 225 million dollars.

- A two day storm interrupted electrical service to just fewer than one million people in Illinois and Missouri.
- Utility replaces 1,316 transformers due to the storm.
- Utility replaces 1,550 power poles due to the Act of God.
- Utility replaces 2 million feet of conductors (cable).

- Human effort to restore power in 9 days of outage included 5,300 people: 2,650 linemen and 1,150 tree crews

Four months later:

- Two day storm interrupted electrical service for same utility for 550,000 people.
- Utility replaces 392 power poles.
- Utility replaces 3 million feet of conductors (cable).
- Human effort to restore power in 8 days was 5,400 people.

One month later:

- One day ice storm interrupts power to 350,000 utility customers
- Human effort to restore power included 5,700 people, specifically 4,200 linemen
- 5 day outage restored with help from contractors from 16 other states.

The storm was not nationally regarded as epic, hardly a storm of Katrina, Rita, Wilma, Hugo or Andrew's size or force. The point is the utilities are extraordinarily challenged by ongoing Acts of God and more recently human intervention. In the post deregulated world, inventories of "roles and poles," linemen, trucks and other inventory are kept to a minimum on the transmission and distribution side. There are consolidated real estate footprints that were once distributed strategically and logically near main roads and urban environments for swift service. Fewer yards, trucks, transformers, roles, poles and cable mean less overhead and similar network reliability, but worse dependability. The "buddy system" of replacing parts, equipment and servicemen in an emergency has replaced established protocols with out of region or "just in time" or out of state solutions. Low or no overhead procedures have encroached upon reliability to the user needs. If you think about it, what is the customer's recourse (commercial or residential) if the power is out for 9 days instead of just 1 or 2?

The lessons learned here for scheduled and outages caused by Acts of God or human intervention are to incorporate self help in energy design and implementation as well as corporate diversification or decentralization of personal data.

For companies, governments, hospitals, heath care and entities that need to operate 24x7, remote facilities are appropriate strategic planning.

Human intervention consequences from a utility point of view are fairly similar. Not to minimize the malicious intent and human or emotional by products associated with terrorism, energy fundamentals are fairly simple and empirical. Power is on or it is off regardless of the source. Human intervention, or man made creations including but not limited to:

1. Nuclear facilities
2. Railroads
3. Cars/trucks
4. Aircraft

5. Waste facilities
6. Gas lines

These contribute to utility risk, as does volatile weather and unique Acts of God.

By way of example, events and legislation are local as they relate to goods and services, local impacts today have global implications. We are truly all in this together.

Chemical manufacturing major sites tend to be built on our crowded coastal or river communities. These, like port cities were based on transportation of goods and services. One of our valued gases is ethylene, used for energy creation and production. Of the 175 million tons, it is estimated that almost 50% of production or distribution is in a high risk head zone at 30% in a medium risk flood zone. These sites globally are unlikely to change or move due to expense. In the states, these sites are all in the Gulf of Mexico, Texas, and Louisiana.

The legislative impact locally with global repercussions are directly related to Acts of God and interruption due to the governing body's ability and willingness to recognize the short and long term effects of "clean up" or fix of same.

Chemical manufacturing processes release greenhouse gases, CO_2, Methane, Nitrous Oxide, Peflurocarbon, Sulfur Lexafluride, etc. Having acknowledged that, emission controls in Europe are a flourishing business. According to one source, "Building a Green Economy," some 64,000 die from soot emitted from power plants each year. In 2002, the World Health Organization reported that 70,000 people died in that year in the US due to pollution. That is twice the number of deaths due to auto accidents.

The point is, that the legacy of energy productions are harmful airborne particulants for humans and agriculture. The impact of energy creation to the planet is negative. That is 25% of the world's total. By thinking outside the grid in a decentralized energy creation mode, we will enhance our collective carbon footprint.

When energy creation effort, time and expense outpaces the energy conservation practices and we will capture the hearts and the imagination of the general public globally, we will lessen the impact or avoid the perfect storm of reduced supply and compounding demand.

Abstract of the Energy Benefits for the Emergency Stabilization Act of 2008

As my book goes to print, the glacial wheels of government expedited energy conservation and energy creation incentives to drive the private sector to "do better." The following is a high level summary of the new law, which is favorable in most cases.

This is a partial abstract of benefits of the Emergency Economic Stabilization Act of 2008, which was signed July 29th 2008.

Title I- Energy Production Incentives- Subtitle A- Renewable Energy Incentives Bill HR 1424 ironically was established for provisions of "Mental Health and Substance Related Disorders Benefits" under group health plans.

The most recognizable and public characteristic of the Bill is the term "Trusted Assets Insurance Financing or Trusted Asset Relief Program (TARP)." This is the term that pundits and economic overnight experts are referring to as they articulate various spins on tax payers support for debt.

This Bill is exactly 451 pages of law crafted over days, not weeks. Sounds daunting and a 180 degree swing from US

Treasury Secretary Paulson's original 3 page request of $700 Billion Dollars.

As crafted, the average word count per line is 5-7 words over 24 lines, so when the overnight "experts" (media) wave the document in front of the cameras and spoke about the wooden arrow benefits developed and additional pork in the Bill; realize that 151 of the 451 pages in Bill HR 1424 are allocated towards Title I: "Energy Production Incentives- Subtitles A Renewable Energy Incentives." **Also consider that if it were single spaced; this Bill could have been reduced significantly.**

The focus of the Energy Incentives is to make existing and new energy creation. It is also intended to make conservation improvements more economically viable and to extend the incentive or benefit horizons.

Lobbyists from the leading energy creation and conservation companies were working full time to meet these milestones in a remarkably short amount of time within this Bill. This is certainly self serving for equipment makers, but I believe the public interest was well served in the Bill.

The language of the Bill for energy enhancements are on pages 115-261. That is one third of the copy of HR 1424!

For corporate uses or landlords nationally but in the New York, New Jersey and Connecticut areas, specifically where the cost per KWH exceeds the national average by 100-200%, solutions provided, equipment manufactured, and energy operators have an enhanced value of proposition.

This energy proposition for decentralized or on site power has never been more favorable in the history of energy creation in the US.

Tax incentives ending in January 1, 2009 now end in January 2011. This is still a tight schedule proposition due to the planning and construction that can be protracted in vertical assets in urban environments, but it is significantly better than it was (pg 115, lines 7-14).

This biomass or energy crop language reflects the country's current mood on ethanol. The mood is not favorable (pg 117 lines 18-25). Energy crops will likely hurt the global economy.

The manufactures for hydro speak to the physical and eco challenges of water management for "non hydro dams." The energy credits are now available, but to design and permit such non-energy creating dams and not affect water levels, even with reservoir pumps, are likely to not be realized (pg 118-119, lines 8-22). Furthermore, making energy from rivers is not a new idea. Most of the water ways that make sense are in place already.

"Refined Coal," trash facilities, expansion of the biomass placed in service after December 31, 2008 needs to be completed in 27 months to get credits. This is a very tight but reasonable schedule in the real world (pg 119-120, lines 19-10).

Energy from waves or marine renewable has a horizon of 40 months (pg 120-122). Creating estimates and diversionary structures with the Army Corps of Engineers is no day at the beach. Credits, money, and time are needed. This Bill reflects that and is an awesome move!

Solar energy property benefits negotiated to expire January 1, 2009 are extended to January 1, 2017 (pg 123, lines 1-4). This is appropriate due to the extraordinary long lead time for large 15-150 megawatt PV cell power cells and the scattered solar power distribution is both expensive and protracted. Transmission and distribution linger as challenges. Real estate is required!

Fuel cell property benefits are extended from December 31, 2008 to December 31, 2016 (pg 123, lines 5-7). This is appropriate due to the expense and technology horizons benefits for large scale hydrogen fuels cells to have economic benefits. **The economic benefits are real or 30% from $1000 dollars per KW to $3000 per KW. This will make the turbines or CHP technology more compelling.**

Microturbine property benefits have the same benefits horizon of December 31, 2016 with the new legislation and the same cost benefits of $3,000 dollars per KW (pg 123 lines 8-16).

These credits move the financial benefits models for less challenging installation from 4-5 years to 2-3 years for assets of 10% combined heat and power (CHP) for systems of energy efficiency of 60% (pg 125, lines 4-19) but under 50 megawatts (pg 126, lines 19-24).

The credits for small wind are $500 per ½ KW up to $4,000 for wind turbines of not more than 100 KW through

December 31, 2016, or $1,667 per ½ KW up to $13,333 for multiple units (pg 131, lines 15-20).

Geothermal's existing credits will extend to January 31st 2017 will include "equipment what uses the grounds or ground waters as a thermal energy sources to heat a structure...or to cool structure" (pg 132, line 21-24) for 30% of qualified geothermal properties (pg 136, lines 13-15) and $2000 for qualified properties, and $6,667 for joint properties (pg 136, lines 5-7). This is good news for the western states.

Tax carry forwards have been considered for years in that all personal credits are allowed against regular and alternate minimum taxes will be extended to the following tax year and allotted to the allowable credits (pg 139, lines 1-12).

New Clean Renewable bonds have been legislated by HR 1424 to support:

1) 100% of the available project of expenses incurred by government bodies, public power provides, or cooperative electric companies for 1 or more renewable facilities.
2) The annual credit to any renewable project will be 70% of the amount determined (pg 171, Lines 5-23).

Limits of Bonds are limited to $800,000,000 and to be distributed as follows:

← **33 1/3%- Qualified public power providers (pg 142, lines 9-11).**

← **33 1/3%- Qualified government bodies (state or tribal) (pg 142, lines 12-14).**

← **33 1/3%- Qualified cooperative project to electronic companies (or not from project electrically) (pg 142, lines 15-17). This is badly needed and good news!**

The Duration of the Clean Renewable Energy Bonds as amended by striking the expiration of December 31, 2008 to December 21, 2009 (pg 145, lines 22-25).

Qualifying FERC or states are:

- Transmitting utility as per Federal Power Act (pg 152 lines 22-24).
- Electric Utility as per Federal Power Act (pg 153, lines 1-5).
- Term expiration is 4 years following the close of the taxable year of the transaction.
- Exceptions are properties outside the USA.

Subtitle B- Carbon Mitigation and Coal provisions (expansion and modification of advanced coal project investment credit) (pg 154).

- Benefits are extended to 30% for qualified projects (pg 154, lines 18-21).
- New expansion of aggregated credits from $1,300,000,000 to $2,550,000,000 dollars.
- The Secretary has allocated $800,000,000 dollars for gas and combined cycle projects.

- The Secretary has authorized $500,000,000 for projects which offer advanced coal based generation technologies (pg 155, lines 6-14).
- The Secretary has authorized $1,250,00,000 for advanced coal based technologies for different durations.
- Application duration- Three years from the date prescribed by the Secretary (pg 156, lines 9-14). Assets applying will include equipment which separates and sequesters 65-70% of the total carbon emissions (Highest separation of carbon gets highest priority) (pg 157, lines 1-5).

Section 112- Expansion and Modification of Coal Gasification Investment Credits

- 30% Credits now exceed 20% credits.
- 350,000,000 Available.
- 200,000,000 in addition for quality gasification projects of 75% or greater of sequestration (pg 160 lines 1-6)

Section 113- Temporary Increase of Coal Excise Tax Funding of Black Lung Disease Fund:

- Extended from January 1, 2014 to December 31, 2018 (pg 161, lines 17-23). Privates funding for those suffering from Black Lung related to the US Government and debt tied to the 1 year Treasury Rate (pg 113, lines 16-23).

Section 114- Special rules for reformed of the coal excise tax to certain producers and experts.

The Law captures a tax refund for coal producers as associated party's having paid excise taxes from October 1, 1990 estimated or equal to $0.825 per ton of coal (pg 168, lines 3-19).

Section 115- Tax Credit for Carbon Dioxide Sequestration.

Now baseline of CO 2's sequestration is $20 dollars per metric ton. This is appropriate due to the 10-20% unique expense to install, retrieve and secure CO2's but also the 10-15% inefficiency that follows the sequestration process for coal and $10 dollars per metric tons for CO2 secured from oil or natural gas (pg 175, lines 5-18).

Criteria for both benefits include but are not limited to assets owed by tax payers, sequestration equipment, equipment and assets that capture no less that 500,000 tons per year (pg 176, lines 11-17).

Some of the new guidelines and extended windows for "Energy Efficient Commercial Buildings" are in section 303 (pg 218, lines 5-18) and specifies parameters for domestic appliance to "do better" in energy consumption and water usage (section 305, pg 218, line 222). They will rise from $75 dollars in 2008 to $200 dollars in 2010. These are all good measures to apply against a set a side amount of $75,000,000 dollars to financially support the changing of equipment for domestic and commercial everyday usage.

As mentioned early in the book, energy conservation measures will provide the faster "bang for the buck" in the short term but energy creation efforts and decentralized self help as well as government subsidizes solutions are the "special sauce" to conquer energy challenges nationally and globally.

Laws relating to steel credits and other were intentionally omitted. The focus here was simply to articulate the new law and the energy conservation/credit benefits.

Reference's/Resources

Department of Energy web site

Department of Defense web site

IPCC- Intergovernmental Panel on Climate change

Patriot Act

Sarbanes Oxley Legislation

Bill S 2191- Energy Act

National Industrialization Protection Plan 2006

Natural Clean Air Act- 2002

Natural Clear Sky Amendment- 2004

Kyoto Protocol

Nuclear Regulatory Committee web site

National Energy Resource Center (NERC) web site

Regional Greenhouse Gas Institute (RGGI) web site

NYSERDA web site

NY State Public Service Mission web site (PSC)

Environmental protection agency (EPA) web site

Department of Environmental Protection (DEP) web site

Green Grid

Congressional Budget office (CBO) web site

Chicago Carbon Exchange web site

ECX- European Carbon Exchange web site

Clean Air Interstate Rule (CAIR)

American Wind Energy Association

Power Utility Commission (PUC)

National Geographic Magazine

Newsweek Magazine

Business Week Magazine

Mother Jones Magazine

Transmission and Distribution World Magazine

Intergovernmental Panel on Climate Change (IPCC)

Carbon discursive Project (CDP)

The Pew centers on Climate Change

The World Business Council Sustainable Business

The New York Times

Platts Guide- McGrawhill- Web site and Maps

Power Magazine

Bob Bodey's various data points and Financial analysis

U.N.-Climate Study

World Metrological Organization

U.N. Millennium Project

The Kyoto Protocol

Rising Tide

IMF World Economic Outlook

Green Peace 2006

USGS

Glossary

Many of the definitions in this glossary are derived from language from Federal Laws, acts, included in national plans including the Homeland Security Act of 2002, Clean Air, Clear Skies, Climate Security Act, the USA Patriot Act of 2001, National Incident Management System, and the National Response Plan, American Climate Security Act, as well as common phrases and jargon from the energy creation, distribution, and mission-critical world.

I have often been accused of having my own hybrid language, one I call a cross between hip-hop and engineering. Friends, clients, and customers often smile and shake their heads when I go off on my techno-babble diatribes.

The world of energy and mission-critical infrastructure—that is, "rocket ship real estate" or facilities—has its unique language relevant to energy creation and conservation, outside plant considerations, inside plant considerations, maintenance, management, and service-level agreements.

You need to use a certain amount of patience to understand the language. Do not push back. Be brave, and embrace some of the language. This whole topic is a lot more interesting.

AC Power:

Alternating current relate to power distribution and condition

Active-Active:

Data center topology for other synchunous real time transfer of Data

Acts of God:

Events control by things other than man make. Often referring to weather or earth conditions of high winds, lightning, earthquake, hurricane, snow fall, rain etc.

Alternative Energy:

Power created or transmitted by new and unusual means. Tradition methods include fossil fuel burning, nuclear or hydro manipulation. Alternative energy would be solar, geothermal, wind, tidal, wave, bio mass, bio waste etc

Also Ran:

Jargon referring to almost legitimate or nearly there.

American Climate Security Act 2007:

Sets caps and goals for Greenhouse gas emissions for energy generation. Frames cap and trade programs for EPA to execute.

Anomolic:

Unusual or unplanned event

Asset.

Contracts, facilities, property, and electronic and nonelectronic records and documents. Unobligated or unexpected balances of appropriations in funds or resources.

America's Climate Security Act 2007:

Law sets limit or cap of certain greenhouses gases emitted from electric generating companies. Law establishes cap and trade templates for Greenhouse Gases (GHG's) and Hydro Fluoro Carbon (HFC's)

Ampere:

Electrical unit of measure.

Backup generators.:

A methodology for creating or storing backup files. The youngest or most recent file is referred to as the son, the prior file is called the father, and the file of two generations older is the grandfather. This backup methodology is frequently used to refer to master files or financial applications.

BANANA:

Build Absolutely Nothing Anywhere Near Anything

Bio-mass:

Energy created from the burning of organic materials. Wood, switch grass, Jatropa.

Bio Digester Energy:

Energy created by the decomposition and fermentation of organic materials or waste products

Bowmanese:

Uniquely crafted language or terms. Origin is Ron Bowman.

Business continuity:

The ability of an organization to continue to function before, during, and after a disaster.

Business impact analysis (BIA):

The process of identifying the potential impact of uncontrolled, nonspecific events in an institution's business process.

CAIR:

Clean Air Interstate Rule

Canola:

Edible oil derived from plants

Carbon Capture and Sequestration (CCS):

Pre or Post energy creation (combustion) of harmful Greenhouse Gases

Carbon Credit:

Unit of measure by the ton of paticulants place in the atmosphere

Carbon Neutral Energy:

Energy created without the creation and distribution of particulants in the atmosphere that impact greenhouse gases, or energy that has a zero sum equation of contribution of creation minus the addition of paticulants inhibitors (like trees).

CBO:

Congressional Budget Office

CCX:

Chicago Carbon Exchange

Centralized Power:

Power from a generally large plant for massive distribution to multiple transmission lines for further distribution.

Centrex Telcom:

Traditional telecom addressable system for voice systems.

CH4:

Methane gas compound

Chindia:

China and India- Jargon for the collective needs or consequence for two high growth, high need, very productive parts of the world

CHP:

Combined heat and power. Generally refers to cogeneration systems.

Chernobyl Meltdown:

1986 catastrophic nuclear power plant explosion in the former USSR where 4,000 people died and others injured.

Clean Air Act of 2002:

Amends the Title 4 of Clean Air and establishes a cap and trade program requiring reductions of CO_2 and other particulants.

Clean Coal:

Coal that has been manipulated to extract the dirty components or particulants of Sulfur and other

CLEC:

Competitive- local Exchange Companies

Clear Skies Amendment 2003–4:

Provides more guide lines for cap and trade and carbon sequestration goals.

CO2:

Carbon dioxide

Cogeneration:

Invented by Thomas Edison, the capture of the waste product of energy creation, heat and making use of the energy waste product for cooling or additional heat.

Consultant

New York for unemployed. Sometimes a value added person or company service.

Control systems:

Computer-based systems used within many infrastructures and industries to monitor and control sensitive processes and physical functions. These systems typically collect measurements and operational data from the field, process and display the information, and relay control commands to local or remote equipment or human/ machine interfaces (operations). Examples of types of control systems include SCADA systems, process control systems, and digital control systems.

Coopetition

The cooperation and competition of companies by sharing ideas or best practices

Corn:

Agricultural product for human and animal consumption. Base product for other sugar and corn syrup products.

Critical infrastructure:

Asset systems and networks, whether physical or virtual so virtual to the United States that the incapacity or destruction of such assets, systems, or networks would have debilitating impact on security, national economy, economic security, public health or safety, or any combination of those.

Critical financial markets.:

Financial markets whose operations are critical to the U.S. economy, including markets for Federal Reserve funds, foreign exchange, commercial paper, and government, corporate, and mortgage-backed securities.

Critical task:

Those prevention, protection, response, and recovery tasks that require coordination among the appropriate combination of federal, state, local, tribal, private sector, or nongovernmental entities during major events in order to minimize the impacts on lives, property, and economy.

Cybersecurity:

The prevention of damage due to unauthorized use or exploitation of, and, if needed, the restoration of, electronic information and communication systems, and information contained therein to ensure confidentiality, integrity, and availability. It includes the protection and restoration, when needed, of information networks and wireless satellite public safety answering points, 911, 411, communication systems, and control systems.

Dam:

Method of holding of retaining water en mass for future consumption, energy creation or irrigation.

Data synchronization:

The comparison and reconsolidation of interdependent data files at the same time so they can contain the same information.

DC Power:

Direct Current

Decentralized Power:

Energy created closer to the usage and away from traditional or incumbent creation.

DEP:

Department of Environmental Protection

Dependency:

The one-directional reliance of an asset system network or a collection thereof within or across such sectors or input, interaction, or other requirements from other sources to function properly.

Digester Gas:

From anaerobic decomposition for organic materials. The harvesting of gases from waste and decomposition. In these purposes to manage gas creation and usage.

Disaster recovery plan:

A plan that describes the process to recover from major unplanned interruptions.

Distribution Power:

From energy creation the transmission of power over conductors or circuits for usage over generally shorter distances. Transmission implies greater distances.

DOE:

Department of Energy

EBIDA:

Earning before interest, depreciation and amortization

Emergency:

An occasion or incidence for which in determination of the present federal assistance is needed to supplement state and local efforts and capabilities to save lives, and to protect property and public health and safety to lessen or avert the threat of catastrophe in any part of the United States.

Emergency response provider:

Includes federal, state, local, tribal agencies, public safety, law enforcement, emergency response, emergency medical including hospital emergency facilities, that related to personnel agencies and authorities (see section 2.6, Homeland Security Act 2002).

Emististic Glass:

Silver ionization of particles fixed to glass to absorb and retard destructive rays of the sun.

Encryption:

The conversion of information to code or cipher.

Energy Darwinism:

Power creation or distribution life cycle analysis that considers and incorporates survivability of practices technologically and financially.

Energy Deregulation 2002:

Privatized or decentralized power creation and distribution in the United States. Effectively created a power generating vehicle and transmission or wire services vehicle of power creation and distribution.

Energy Star:

Rating created by the Department of energy to meet certain criteria of energy conservation or creation to benefit the environment

EPA:

Environmental protection agency

EPS (emergency power systems):

A generator-only method of creating and storing power.

Ethanol:

Ethyl alcohol or grain alcohol

Euclidean Miles:

Point to Pont. The shortest distance.

Everything is on the Table:

Jargon implies that under dire circumstance most reasonable ideas will be considered.

EV-I:

All electric car make by GM. Very successful.

FEMA:

Acronym for Federal Emergency Management Agency.

Fuel oil:

Fossil or organic energy source.

Farmageddon:

The idea that the farming or agricultural ideas and current implementation will be more harmful than helpful for various reasons

Farmatopia:

The idea that the farming of agricultural ideas and current implementation will be more helpful than harmful for various reasons.

FIP:

Federal Implemental Plan

First responder.:

A local and nongovernmental police, fire, and emergency person, who in the early stages of an incident is responsible for the protection and preservation of life, property, evidence, and the environment, including emergency response providers as defined in section two of the Homeland Security Act of 2002, as well as the emergency management, public health, clinical care, public works, and other skilled support personnel. Jurisdiction, a range or sphere of authority public agencies have jurisdiction on and incident related to their legal responsibilities and authority. Jurisdictional authority at an incident can be "geographical."

Fossil Energy:

Energy created by hundreds of millions of years of organic waste, time and pressure have created solid or liquid flammable or combustible energy.

Fur Ball:

Vernacular IT expression implying multiple hardware and software configuration compiled over time and not efficient.

Gap analysis.:

Comparison that identifies the difference between actual and desired outcomes. Generally accepted practices.

Geothermal Energy:

Energy created by harnessing the earths heat the closer to the vents or earths center to create hot water or steam and move turbine engines.

GDP:

Gross Domestic Product

Gigawatt:

Energy equivalent to one billion watts

Global Warming:

The idea or science that the due to ozone depletion and an abnormal increase of particulants and CO_2 emissions that the atmosphere is warming causing a host of unanticipated and unwelcome consequences.

Glocally:

Events and processes that have similar effect over a smaller or regional footprint as well as world wide or global impact.

Green Job:

A full time or part time job/career associate with earth enhancing activities. Forestation, bio wastes energy creation, recycle activities, energy conservation activities.

Green Gold:

Value added processes or products that make money.

Green Grid:

Group of computer and technology companies in association to create energy conservation methods of chip creation and utilization to benefit the end user and environment

Green movement:

Social and economic concentrated involvement and manipulation of earth friendly ideas and businesses. Energy conservation, environmentally saving and preservation concepts are cherished.

Grid:

A network of power generating plants, substations, transmission lines and distribution lines.

GTL:

Gas to liquid technology

HAZMATS:

Hazardous materials

Heat Island:

Urban environment or cities that are warmer due to density of people, buildings, transportation, and equipment, without landscape or foliage to cool.

HIPPA:

The American Health Insurance Portability and Accountability Act of 1996. A set of rules for health care and physicians to follow to protect users rights and record management.

Hockey Stick:

Refers to unusually high growth in the shape of the head of a stick

Human Intervention:

Willful or unwillful act of a person or persons. Often referred to in terrorism and catastrophic events

Hybrid Car:

Car or truck using battery technology currently to offset gas or liquid petroleum usage in idling stage of transportation

Hydro Energy:

Water created power via dam, tidal, wave bob or other manipulation.

Hydrogen Fuel Cell:

Electrochemical conversion devise. It produces electricity from fuel on anode side and an oxidant on the cathode side that reacts in the presence of an electrolyte. Sources of fuel are natural gas and water manipulation.

HVAC:

Acronym for heating, ventilation, and air conditioning.

ISP:

Inside Plant. Service inside a facility often mechanical, electrical and plumbing as well as telecom.

Infrastructure:

The framework of interdependent networks and systems comprised in identifiable industries, institutions (including people and procedures), and distribution of capabilities that provide the reliable flow of products and services.

IPP:

Independent power provider

Jatropa:

Plant grown and rich in oil in large quantities (hectars) in often dry or non agriculturally friendly parts of the world. A source of energy. Low methane and low impact on agricultural needs for food substance.

Just in time-power:

Power that is provisioned to burn at the exact moment it is needed based on real time and historic information.

Key resources:

Publicly or privately controlled resources essential to the minimal operations to the economy and government.

Kill Zone:

Area of immediate death due to exposure of radio active materials

Knuckleheads:

Vernacular Jargon implying incompetence, stupidity or silliness. Refers to or describes underwhelming consults, real estate brokers, or alleged experts in their field.

KVA:

Kilovolt ampere

KW- Kilawatt:

One thousand watts

Kyoto Protocol:

173 nation agreement setting down criteria down timelines for environmentally friendly goals.

LD:

Long distance, in reference to telecom providers

LEED:

Leadership in Environmental Design, criteria of low carbon emitance and resources conservation and creation. Established in 1998.

Lemming:

Jargon referring to user or experts following the next without regarding for relevant intelligence.

LMOP:

Landfill methane outreach program

Major disaster:

A natural catastrophe (including any hurricane, tornado, storm, high water, wind-driven water, tidal wave, tsunami, earthquake, volcanic eruption, landslide, mudslide, snowstorm, or drought, regardless of cause of any fire, flood, or explosion in any part of the United States), which to the determination of the president causes damage of sufficient severity and magnitude to warrant major disaster assistance under the Robert Stafford Disaster Relief and Emergency Assistance Act to supplement the efforts and available resources of the state and local governments.

Media:

Physical objects, stored data such as paper, hard disc drives, tapes, and compact discs (CDs).

Megawatts:

One million watts

Methane:

Chemical compound CH4

Micro-dam:

Smaller dam, generally under 5 megawatts of power.

Mirroring:

A process that duplicates data to another location over a computer network in real time or close to real time.

Mission Critical:

Facilities or technology that have loss of life of money emphasis on uptime reliability. 24 × 7 facilities access, design and maintainability.

Mitigation:

Activities designed to reduce or eliminate risk to persons or property or to lessen the actual or potential effects or consequences of an incident. Mitigation measures may be implemented prior to, during, or after an incident. They often are developed in accordance with the lessons learned from prior events. Mitigation involves ongoing actions to reduce exposure to, probability of, or potential loss from hazards.

Moore's Law:

The doubling of data creation and bandwidth every 18 months, put forth by Intel founder Gordon Moore.

Moratornm or uncontrolled power plant act of 2008:

Will require new or retrofit fossil fuel power plants to sequester 85% of CO_2 emissions.

Network:

The group of assets or systems that share information or interact with each other in order to provide infrastructure services within or across sectors. Preparedness, the range of deliberate critical tasks and activities necessary to build, sustain, improve, and implement operational capability to prevent, protect against, respond to, and recover from domestic incidents.

Negawatt:

The negative absorption of power creation or usage through conservation efforts.

NERC:

National electric regulatory committee

Newbie:

New comer to a field of experts

New Math:

Wall Street Analyst PE rationalization or other metric of absurd or out of market expectations. A black hole of analyst reasons for over initial stock prices.

NFPA:

National Fire Protection Association

NIMBY:

Not in my backyard

NOPE:

Not On Planet Earth- refers to far left interests for not drilling for oil in US

NOX:

Nitrogen Oxide

Nuclear Energy:

Energy created by the fusion or of uranium molecules

NYSEDRA:

New York State Energy Department Research Authority

OPEC:

Organization of Petroleum Exporting Countries

OSP:

Outside Plant. Services outside a faculty- often utilities and telecom.

Overnight experts:

Consults or Brokers often equipped with 3–5 handy acronyms with perceived value added finger tip knowledge. Reality is often after you scratch the surface of their knowledge, there is more surface.

PBX:

Acronym for private branch exchange.

PE Ratio:

Prince per earns up ratio

Perfect storm:

Relating to Energy- the near term forecast of Demand outpacing supply considering ongoing conservation practices, generating sources and decommission of legacy assets.

PPA's:

Pacific Power Association

PMU:

Power management unit

Preparedness:

A continuous process involving efforts at all levels of government and between government and the private sector and nongovernmental organizations to identify threats, determine vulnerabilities, and identify required activities and resources to mitigate risk.

Prevention:

Actions taken to avoid an incident or to intervene to stop an incident from occurring. Prevention involves actions taken to protect lives and property. It involves applying intelligence and other information to a range of activities that may include some countermeasures, such as deterrence operations, heightened inspections, improved surveillance, security operations, and investigations to determine the full nature and source of that threat.

Prioritization:

The process of using risk assessment results to identify where risk reduction mitigation efforts are most needed and subsequently to determine which protective actions should be instituted in order to have the greatest effects.

Protection:

Actions to mitigate the overall risk to critical infrastructure and key resources (CI/KR) assets, systems, networks, or their interconnecting links resulting from exposure, injury, destruction, incapacitation, or exploitation. Protection includes actions to deter the threat, mitigate vulnerabilities, or minimize consequences associated with a terrorist attack or other incident. Protection can include a wide range of activities, such as critical facilities, building resiliency and redundancy, incorporating hazard resistance into initial facility design, initiating activity or passive countermeasures, installing security systems, promoting workforce security, or implementing cybersecurity measures among various others.

Power Network:

The infrastructure of energy generation, transmission and distribution.

PSC:

Public Service commission- New York State

Public and private sector entities:

Often quote risk management frameworks in their business continuity plans.

PUC:

Public Utility commission

RBOC:

Regional bell operating company for Telcom providers

Recovery point objectives (RPOs):

The amount of data that can be lost without severely impacting the recovery of operations.

Recovery site:

An alternative location for processing information (and possibly conducting business in an emergency). Usually divided between hot sites, which are fully configured centers with compatible computer equipment, and cold sites, which are operational computer centers without the computer equipment.

Recovery time objectives (RTOs):

A period of time that a process can be inoperable.

Renewable Energy:

Power or energy for continuous sources like the sun, wind, geothermal and agricultural.

Resiliency:

The capacity of an asset, system, or network to maintain its function during, or to recover from, a terrorist attack or other incident.

Response:

Activities that address the short-term direct effects of an impact or incident, including immediate actions to save lives, protect property, and meet basic human needs. Response also includes the execution of emergency operation plans and migration activities designed to limit the loss of life, personal injury, property damage, or other unfavorable outcomes. As indicated by the situation response activities, response includes applying intelligence or other information to lessen the effects of the consequences of the incident.

RGGI:

Regional Greenhouse Gas Institute

RIF:

Reduction in force

Risk:

A measure of potential harm that encompasses threat, vulnerability, or consequence. Risk is the expected magnitude of loss due to a terrorist attack, national disaster, or other incident.

Risk management framework:

A planning methodology that outlines the process for setting security goals, identifying assets, systems, networks, and functions, assessing risk, prioritization, and implementation of protective programs, measuring performance, and taking corrective actions.

Rocket ship Real Estate:

Generally mission critical facility that are uniquely improve

Route Miles:

The distance from point to point as traveled. Often indirect. Generally 40–608 greater then Eudiclean miles

Routing:

The process of moving information from its source to its destination.

Salon:

Physical or virtual place where the best and the brightest exchange ideas to discuss and critique.

SAS 70:

An audit report of a servicing organization prepared in accordance with guidance provided by the American Institute of Certified Public Accountants Statement on Auditing Standards Number 70.

Sector:

A logical collection of assets, systems, or networks that provides common function to the economy, government, or society. HSPD 7 defines 17 critical infrastructure key resource sectors.

Server:

A computer or other device that manages a network service. An example of print server device that manages network printing.

SHIT:

Store High in transit. Stamp on Acronym place on dry manure during the early days of shipping.

SOHO:

Small office, home office

Solar Energy:

Energy captured and manipulated as well as stored for future usage.

Solar PV Cell:

Tool of solar energy, photovoltaic cell

Sonet:

Synchronous Optical Network. Implies a self healing and continuous reliability of telecom integrity.

Sarbans Oxley (SOX):

Law enacted for corporate accounting verification with come basic business continuity language for the preservation of data.

SOX:

Sulfur Oxide

Special Sauce

Term used to imply unique or value added process or event by various business or industry.

Subsurface Geology:

Study of the various levels of the earth and resources.

Sugar Cane:

Agricultural product planted and harvested for human consumption and energy creation

Sustainable Energy:

Often referred to as renewable energy.

System development life cycle (SDLC):

A written strategy or plan for the development and modification of computer systems, including initial approvals, development documentation, testing plans, results and approvals, and documentation of subsequent modifications.

U:

Unit of measure-for cabinets within data center

T1 Line:

A specific type of telephone line for digital communication only.

TCO:

Total cost of ownership

Tariffs Energy:

Governments proportion of income of expense of a product or service.

Terrorism:

Any activity that (1) involves an act that is dangerous to human life or potentially destructive of critical infrastructure or key resources, and a violation of criminal laws in the United States or of any state or subdivision of the United States, and (2) appears to be intended to intimidate or coerce the civilian population, influence the policy of a government by intimidation or coercion, or affect the conduct of a government by mass destruction, assassination, or kidnapping.

Thermal Storage:

The housing of chilled slurry or water for cooling benefits when fans are employed to blow. Same principles are employed for contained hot water.

Threat:

The intention and capability of an adversary to undertake actions that would be detrimental to the critical infrastructure and key resources.

Three Mile Island Nuclear Power Facility:

Long island, best know for a partial core melt down beginning on March 28th, 1979. 25,000 people lived within 5 miles and no cancer evidence or consequences have been reported according to some resources.

Tier:

Groupings of jurisdictions that account for regional differences in expected capability levels among entities based on assessments of total population, population density, critical infrastructure, and other significant factors.

Tidal Energy:

Energy harnessed from the hydro flow of the oceans tides in and out of captured and contained areas. Tidal power can be harnessed by fixed earth mounted turbines as well.

Transmission Power:

Energy transported in generally large sized or qualities for generating station to substations.

Tree Hugger:

Jargon referring to earth friendly or eco friendly person.

Tree Trimming

Maintenance critical for utility transmission and distribution. Fallen braches or trees is a large source of foul weather outages in Power network

TVA:

Tennessee Valley Authority. Eight State Power Utility.

Two Point Five namseter (2.5 mn)

Partuants- the grouping of six Greenhouse gases of unique interest.

UPS (uninterruptible power supply):

Typically a collection of batteries and power rectifiers that provide electrical power for a limited period of time.

Uranium:

Silver –grey metallic chemical. Periodic table symbol U. Has 92 protons and 92 electrons.

Vaulting:

A process that periodically writes backup information over a computer network directly into a recovery site.

VAR:

Value added reseller (resource)

Vulnerability:

A weakness in the design, implementation, or cooperation of an asset, system, or network that can be exploited by an adversary or destroyed by a national hazard or technological failure.

Wave Energy:

Energy or power created and stored by movement of water/waves and the inertia is captured. Wave bobs are commercially deployed today globally.

Weapons of mass destruction:

Any explosive, incendiary, or poisonous gas, bomb, grenade, rocket having a propellant charge of more than four ounces. Missile having an explosive or incendiary charge or more than one-quarter ounce, or mines. WMD are also any weapons that are designated or intended to cause death or serious bodily injury through the release, decimation, or impact of toxic poison chemicals. These include any weapons involving a disease organism, radiation or radioactivity at a level that is dangerous to human life.

Wikinomics:

"Fast" economics. The speed of business.

Wind Energy:

Energy created and store by the manipulation and storage of wind movement over blades to move turbines or generators. Energy is transmitted or stored

Yucca Mountain

Nuclear waste repository in Nevada, US.

Index

Ronald H. Bowman, Jr.

57 Finley Road
Princeton NJ 08540

Ronbowman123@aol.com

Professional Career:
Mr. Bowman is a veteran
infrastructure and real
estate professional having
over 25 years of experience
in directing leasing,
consulting, construction,
and management programs,
both nationally and
internationally.

Prior to joining Tishman
Technologies, Mr. Bowman founded the Unique
Infrastructure Group, LLC, which provided specialty
consulting and real estate advisory services for multi-
market, uptime, critical data and telecom facilities
nationally.

Mr. Bowman has extensive international experience in
mission critical and infrastructure consulting having

served as a lead consultant on trading floor and data center installations in some of the world's most competitive markets, including London, Johannesburg, Tokyo, Frankfurt, Paris, and Shanghai, among others.

Media: Mr. Bowman is a recognized industry leader, who has been quoted in the Wall Street Journal, The New York Times, USA Today, Crain's New York Business, Real Estate Weekly, Real Estate Journal, Tri-State Real Journal, Engineering Today, Office and Industrial Properties, Star-Ledger, The New York Post, and The New York Daily News.

Regarding intelligent buildings, Mr. Bowman has been interviewed on The Today Show, New York One, Live at Five New Jersey Network, and News 4 New York.

He has given radio interviews for Bloomberg Radio and 1010 Wins. He is regularly called upon to give speeches, and has spoken at such events as: NACORE, Building Seminar, Baruch College, French Intelligent Building Institute, Japanese's Intelligent Building Consortium, Communications Deregulation Seminar, Physical Infrastructure Symposium, Future of Wireless Communication in New York, Corporate Real Estate Conference, UBS Technical Conference, Wall Street Security Forum, Building Wireless Conference, Texas Economic Forum, 24 X7 Mission Critical Conference, and numerous broker/owner presentations on intelligent and futuristic buildings, are we safer post 9/11 and the future of Energy.

Author: He is the author of the recently published book, *Business Continuity Planning for Data Centers and Systems: A strategic Implementation Guide*. He is the author of over 20 White Papers and by line articles.

Personal: Ronald Bowman is an accomplished athlete. He is a three time Ironman finisher. He has competed in the father/son Ironman as well as multiple half Ironman distance races. He is an Olympic distance tri-athlete who has qualified for national championships for the past three years at half and Olympic distances. He has competed in the Boston, Miami, Nova Scotia, and New Jersey marathons. He has completed the New York Marathon three times and has been age group medal winner for the 5K, 10K, and half marathon distances. He has swum around Manhattan twice on relays; six man and two man. Ron and his son, Connor made up the two man relay around Manhattan and was the only father son team to do so. He has participated in a six mile race in the Hudson, a 5 mile race in St Croix, as well as multiple open water races from 1 to 6 miles. He has also won medals in various road races and triathlons, and races with his wife, Maureen, and his daughter, Ceara.

He Co-founded *Impact USA*, an Inner city sports program. He is an active NYC coalition of the Homeless leader by collecting clothes, cooking, feeding, and sleeping in shelters with NYC's homeless in Grand Central, Penn Station, and St. Barts Church.